THE OPEN MAPPING AND CLOSED GRAPH THEOREMS IN TOPOLOGICAL VECTOR SPACES

BY

TAQDIR HUSAIN

McMaster University

SPRINGER FACHMEDIEN WIESBADEN GMBH

1965

Diese Arbeit wurde in Rahmen des Jubiläumspreisausschreibens des Verlages Friedr. Vieweg & Sohn mit einem Preis ausgezeichnet. Die Übersetzung in die englische Sprache besorgte der Verfasser

This work won a prize in the Jubilee Prize Competition held by the Verlag Friedr. Vieweg & Sohn. The English translation has been made by the author

ISBN 978-3-322-96077-1 ISBN 978-3-322-96210-2 (eBook)
DOI 10.1007/978-3-322-96210-2

Originally published by *Friedr. Vieweg & Sohn, Braunschweig in 1965*

TO

MARTHA

PREFACE

THE main purpose of writing this monograph is to give a picture of the progress made in recent years in understanding three of the deepest results of Functional Analysis—namely, the open-mapping and closed-graph theorems, and the so-called Krein–Šmulian theorem.

In order to facilitate the reading of this book, some of the important notions and well-known results about topological and vector spaces have been collected in Chapter 1. The proofs of these results are omitted for the reason that they are easily available in any standard book on topology and vector spaces e.g. Bourbaki [2], Kelley [18], or Köthe [22]. The results of Chapter 2 are supposed to be well known for a study of topological vector spaces as well. Most of the definitions and notations of Chapter 2 are taken from Bourbaki's books [3] and [4] with some trimming and pruning here and there. Keeping the purpose of this book in mind, the presentation of the material is effected to give a quick résumé of the results and the ideas very commonly used in this field, sacrificing the generality of some theorems for which one may consult other books, e.g. [3], [4], and [22].

From Chapter 3 onward, a detailed study of the open-mapping and closed-graph theorems as well as the Krein–Šmulian theorem has been carried out. For the arrangement of the contents of Chapters 3 to 7, see the Historical Notes (Chapter 8).

The bibliography, which can be regarded by no means as complete, is given at the end of the Historical Notes. For detailed references one can consult [7], besides other books referred to above. Square brackets containing numerals indicate references in the bibliography. The reference of definitions of terms used in this book can be found in the index which is preceded by an index for notations and symbols.

The author is greatly indebted to his wife, Martha Husain, for her devoted co-operation and assistance in preparing this book, especially in rendering the German translation of this book which, under the title *On Topological Vector Spaces with Emphasis on the Open Mapping and Closed Graph Theorems*, won a prize in the Jubilee Competition sponsored by the Friedr. Vieweg & Sohn Publishing House, Braunschweig, W. Germany, on its 175th anniversary. The author takes great pleasure in thanking the Friedr. Vieweg & Sohn Publishing House for their award and kindness in agreeing to publish the book in co-operation with the Clarendon Press, Oxford. My thanks are also due to the editors and the staff of the Clarendon Press for producing this book so carefully.

TAQDIR HUSAIN

McMaster University
Hamilton, Canada

CONTENTS

1

ELEMENTARY CONCEPTS CONCERNING TOPOLOGICAL AND VECTOR SPACES

1. Definition of a topological space

LET X be a given set and $u = \{U\}$ a family of subsets of X such that:

(a) X and the empty set $\emptyset$ belong to u.
(b) The intersection of any finite number of sets of u belongs to u.
(c) The union of any arbitrary number of sets of u is also in u.

Then X with the family u is called a *topological space* and denoted by X_u. The elements of X are called *points* of X. The members U of the family u are called *open* sets of X in the topology defined by the family u (or, in short, *u-open*). u is called a *topology* on X.

A subfamily $u_0 = \{U_\beta\}$ of $u = \{U\}$ is said to be a *basis* of the topology u on X, if each u-open set is the union of sets in u_0. Also a subfamily u_0 is called a *sub-basis* if finite intersections of sets in u_0 form a basis of the topology u on X.

A topological space X_u is said to satisfy the *second axiom* of *countability* if u has a countable basis.

If there are two topologies u and v on X such that every v-open set is u-open, then u is said to be *finer* (or *stronger*) than v and denoted by $u \supset v$. If u is finer than v, then v is *coarser* than u, i.e. $v \subset u$. In other words, '$u \supset v$' and '$v \subset u$' are equivalent. $u = v$ if and only if $u \supset v$ and $u \subset v$.

A topology u on the set X is called *Hausdorff* (or *separated*) if, for each pair of distinct points x_1 and x_2 in X, there exist two disjoint u-open sets U_1 and U_2 such that $x_1 \in U_1$ and $x_2 \in U_2$.

A set C in X is said to be *u-closed* if the complement $X \setminus C$ of C in X is u-open.

The family $\{C\}$ of all closed sets of a topological space X_u satisfies the following properties:

(a′) X and $\emptyset$ are u-closed.
(b′) The union of any finite number of u-closed sets is u-closed.
(c′) The intersection of any arbitrary number of u-closed sets is u-closed.

The *closure* $\bar{A}$ of a set A in a topological space X_u is the smallest closed set in X_u containing A. Since the intersection of closed sets is closed, $\bar{A}$ is closed. Further, a set A is closed if and only if $\bar{A} = A$. If we wish to emphasize the topology u in which the closure is taken we shall denote $\bar{A}$ by uA.

Let A and B be two subsets of a topological space X_u. A is said to be *dense* in B, if $\bar{A} = B$. A topological space X_u is said to be *separable* if there exists a countable dense subset of X.

If Y is a subset of a topological space X_u, then Y can also be topologized as follows: For each U in u, let $U \cap Y$ be an open set of Y. It is easy to see that the collection of sets $U \cap Y$, U running over the family of sets in u, defines a topology on Y. Such a topology is called the *relative* (or *induced*) topology of Y.

2. Bases and sub-bases of a neighbourhood system

Let X_u be a topological space. A set V is said to be a *neighbourhood* (*u-neighbourhood*) of a point $x \in X$ if there exists a set U of the family u such that

$$x \in U \subset V.$$

In other words, a set V containing a point x is said to be a neighbourhood of x if there exists a u-open set U containing x and contained in V.

Clearly a non-empty open set P is a neighbourhood of each of its points. Furthermore, a set P is open if and only if it contains a neighbourhood of each of its points ([18], p. 38, Theorem 1).

Let $\mathscr{U}_x$ denote the system of all neighbourhoods of a point $x \in X_u$. Then $\mathscr{U}_x$ satisfies the following properties:

(n_1) $\mathscr{U}_x$ is non-empty and $x \in \bigcap U_\alpha$, $U_\alpha \in \mathscr{U}_x$.

(n_2) If U is in $\mathscr{U}_x$ and $U \subset W$, for any set W in X, then W is in $\mathscr{U}_x$.

(n_3) Each finite intersection of sets in $\mathscr{U}_x$ belongs to $\mathscr{U}_x$.

(n_4) If $U \in \mathscr{U}_x$, then there is a $V \in \mathscr{U}_x$ such that $V \subset U$ and $U \in \mathscr{U}_y$ (where $\mathscr{U}_y$ is the system of all neighbourhoods of a point $y \in V$) for each $y \in V$.

Conversely, given the system of neighbourhoods of each point x of a set X, satisfying the above properties from (n_1) to (n_4), then one can define a topology u on X such that the given neighbourhood system of each point is precisely the neighbourhood system with respect to u. For details see p. 56 of *General Topology* by J. L. Kelley [18].

Let $\mathscr{U}_x$ denote the neighbourhood system of a point $x \in X_u$. A subfamily $\mathscr{V}_x$ of $\mathscr{U}_x$ is said to be a *basis* or a *fundamental system* of *neighbourhoods* of x if, for each $U \in \mathscr{U}_x$, there exists a $V \in \mathscr{V}_x$ such that $V \subset U$.

A topological space X_u is said to satisfy the *first axiom of countability*, if each point $x \in X_u$ has a countable fundamental system of neighbourhoods.

If a topological space X_u satisfies the second axiom of countability, then it also satisfies the first axiom of countability. But the converse is not true.

If a topological space X_u satisfies the second axiom of countability then it is separable. However, the converse is not true. For metric spaces (see § 3) the converse is true.

3. Metric spaces

A set E is said to be a *metric space* if there exists a real valued function d defined for each ordered pair (x, y), where $x, y \in E$ and such that

(m_1) $d(x,y) \geqslant 0$, and $d(x,y) = 0$ if and only if $x = y$.

(m_2) $d(x,y) = d(y,x)$.

(m_3) $d(x,z) \leqslant d(x,y)+d(y,z)$ $(x, y, z \in E)$.

The set of all x in a metric space E such that $d(x, x_0) < r$ $(\leqslant r)$, for a fixed point $x_0 \in E$ and a fixed positive real number r, is called an *open* (*closed*) *ball* of radius r and centre x_0. The totality of all open balls, when r runs over the set of rational numbers, gives a fundamental system of neighbourhoods of x_0. Therefore, in a metric space, each point has a countable fundamental system of neighbourhoods. In other words, each metric space is a topological space and satisfies the first axiom of countability.

A topological space E_u is said to be *metrizable* if E is a metric space such that the system of all open balls defines the original topology u.

A metric space is Hausdorff. Actually it is a *normal space* (i.e. for any two disjoint closed sets C_1 and C_2 there exist two disjoint open sets U_1 and U_2 such that $C_1 \subset U_1$ and $C_2 \subset U_2$) ([18], p. 120, Theorem 10).

A metric space X_u is separable if and only if the topology u has a countable basis ([18], p. 120, Theorem 11).

A sequence x_n $(n \geqslant 1)$ in a metric space is said to be a *Cauchy sequence* if, given $\epsilon > 0$, there exists a positive integer n_0 depending on $\epsilon \ni d(x_m, x_n) < \epsilon$ for all $n, m \geqslant n_0$.

A sequence x_n $(n \geqslant 1)$ in a metric space is said to *converge* to x_0 if, given $\epsilon > 0$, there exists a positive integer $n_0(\epsilon)$ such that $d(x_n, x_0) < \epsilon$ for all $n \geqslant n_0$. Clearly a convergent sequence is a Cauchy sequence. But the converse is not true.

4. Filters and compact sets

Let X be a given set and $\mathscr{K} = \{F_\alpha\}$ a non-empty family of subsets of X. $\mathscr{K}$ is said to be a *filter* on X if the following conditions are satisfied:

(1) Each subset of X containing any F_α in $\mathscr{K}$ belongs to $\mathscr{K}$.
(2) The intersection of any finite number of F_α's belongs to $\mathscr{K}$.
(3) The empty set $\emptyset$ does not belong to $\mathscr{K}$.

A non-empty subfamily $\mathscr{G} = \{G_\alpha\}$ of a filter $\mathscr{K} = \{F_\alpha\}$ is called a *basis* of the filter $\mathscr{K}$ on X if the following conditions are satisfied:

(1′) The intersection of any two sets in $\mathscr{G}$ contains a set of $\mathscr{G}$.
(2′) The empty set does not belong to $\mathscr{G}$.
(3′) Each F_α contains a G_α.

Let X_u be a topological space and $\mathscr{K} = \{F_\alpha\}$ a filter on X_u. $\mathscr{K}$ is said to *converge* to a point x_0 if each neighbourhood U of x_0 contains an F_α.

Taking all systems of filters and ordering them by inclusion, one can obtain a partially ordered system of filters. By Zorn's lemma,† one gets a maximal filter. Such a filter is called the *ultra-filter*. In a topological space X_u, an ultra-filter either converges or has no limit point.

A Hausdorff space X_u is said to be *compact* if each filter on X_u has at least one limit point in X.

One can also describe the compactness in any one of the following ways:

(*a*) X_u is compact.
(*b*) Every ultra-filter on X_u is convergent.
(*c*) If $\{C_\alpha\}$ is a family of closed sets in X_u such that $\bigcap C_\alpha = \emptyset$, then there exists a finite subfamily, $C_{\alpha_1}, \ldots, C_{\alpha_n} \ni \bigcap_{i \leqslant n} C_{\alpha_i} = \emptyset$.

(*d*) Let $\{P_\alpha\}$ be any family of open sets of X_u such that each $x \in X_u$ belongs to some P_α, i.e. $\bigcup P_\alpha = X$. Then there exists a finite subfamily $P_{\alpha_1}, \ldots, P_{\alpha_n} \ni \bigcup_{i \leqslant n} P_{\alpha_i} = X$.

A subset Y of a topological space X_u is a *compact set* if Y, in the relative topology, is a compact topological space.

In a Hausdorff topological space, each compact set is closed. A closed subset of a compact set is compact. The continuous image of a compact set is compact.

† This states that every partially ordered set which is inductive has a maximal element [18].

A subset Y of a topological space X_u is said to be *relatively compact* if $\overline{Y}$ is compact. Every subset of a compact set is relatively compact.

5. Uniform spaces and completeness

Let X be a set and $X \times X$ denote the set of all ordered pairs (x, y), x and $y \in X$. If V is the set of all pairs (x, y), then V^{-1} is the set of all pairs (y, x). Moreover, VU denotes the set of all pairs (x, z), if there exists a y such that $(x, y) \in V$ and $(y, z) \in U$. In VU, one can replace U by V and thus define $V^2 = VV$. The set of all (x, x) $(x \in X)$ is called the *diagonal* set Δ.

X is said to be a *uniform space* if there is a filter $\mathscr{K}$ on $X \times X$ satisfying the following properties:

(1°) Each U in $\mathscr{K}$ contains the diagonal set Δ.
(2°) If U is in $\mathscr{K}$ then U^{-1} is in $\mathscr{K}$.
(3°) For each U in $\mathscr{K}$, there exists a V in $\mathscr{K}$ such that $V^2 \subset U$.

The filter $\mathscr{K}$ satisfying (1°) to (3°) is called a *uniformity* for the set X.

The conditions (1°) to (3°) are generalizations of the properties which define a metric space (§ 3). Clearly a metric space is a uniform space.

A uniformity of a set X defines a topology on X and thus a uniform space becomes a topological space ([18], p. 178).

A uniform space is Hausdorff if $\bigcap U = \Delta$ where $U \in \mathscr{K}$, the uniformity. Each Hausdorff uniform space is *regular* (i.e. for each point x and each closed set C such that $x \notin C$, there exist disjoint open sets U_1 and U_2 such that $x \in U_1$ and $C \subset U_2$) and also *completely regular* (i.e. for each point x_0 and a closed set C such that $x_0 \notin C$, there exists a continuous real valued function $f(x)$ such that $f(x_0) = 0$ and $f(x) = 1$ on C) ([18], p. 188, Corollary 17).

A filter $\mathscr{G} = \{G_\alpha\}$ on a uniform space X is said to be a *Cauchy-filter* if, for each set U in the uniformity $\mathscr{K}$ of X, there exists a G_α such that $(x, y) \in U$ for all $x, y \in G_\alpha$.

Each convergent filter is a Cauchy-filter. If in a uniform space X, each Cauchy-filter is convergent, then X is said to be *complete*.

A metric space E is complete if each Cauchy sequence converges in E.

For each Hausdorff uniform space E, there exists a unique (up to an isomorphism) complete Hausdorff uniform space $\hat{E}$ in which E is dense. $\hat{E}$ is called the *completion* of E.

In a compact uniform space E, each ultra-filter, being convergent, is a Cauchy-filter. Furthermore, each Cauchy-filter in a compact uniform space is convergent. The latter statement implies that each compact uniform space is complete.

A closed subspace of a complete space is complete, and a complete subspace of a Hausdorff uniform space is closed ([18], p. 192, Theorem 22).

In a Hausdorff uniform space E, a set P is said to be *precompact* if $\hat{P}$ is compact in the induced uniform structure. Every compact set of a uniform space is precompact. In a Hausdorff uniform space E, each relatively compact set is precompact. The converse is not true in general. It is true if E is complete.

6. Nowhere-dense sets and sets of first and second category

A subset Y of a topological space X_u is said to be *nowhere-dense* if the *interior* (i.e. the largest open set contained in a given set) of $\overline{Y}$ is $\emptyset$. In other words, Y is nowhere-dense if and only if $X \setminus \overline{Y}$ is dense in X. A subset Y of X_u is of the *first category* in X if Y is the union of a countable family of nowhere-dense sets. A subset Y of X_u is of the *second category* if it is not of the first category.

One of the most fruitful applications of these terms is due to Baire. His theorem states the following:

THEOREM (Baire): *Every metric complete space is of the second category* ([18], p. 200, Theorem 34).

7. Topological products

Let X_α $(\alpha \in A)$ be an arbitrary family of sets. $X = \prod_\alpha X_\alpha$ denotes the set of all functions $x = (x_\alpha)$ such that $x(\alpha) = x_\alpha \in X_\alpha$ for each α. X is said to be the *Cartesian product* of X_α.

If each X_α is a topological space with a topology u_α, then X can also be topologized as follows: Let $U = \prod_\alpha U_\alpha$, where $U_\alpha = X_\alpha$ for all α except for a finite subset B of A and U_α is a neighbourhood of $x_\alpha \in X_\alpha$ $(\alpha \in B)$ in the given topology u_α on X_α. The system of U is easily seen to form a fundamental system of neighbourhoods for each point x in X, when U_α runs over the neighbourhood system of x_α for each α. Thus one gets a topology u on X which is called the *product topology*. The Cartesian product X, with the product topology u, is called the *topological product* of X_α. The mapping p_α: $X \to X_\alpha$ such that $p_\alpha(x) = x_\alpha$ is called the *projection* of X onto X_α.

If each $X_\alpha = Y$ for each $\alpha \in A$, then X is written as Y^A. If $A = N$ (the set of all positive integers) then Y^N is the set of all sequences of Y.

If A = a finite set of n elements and $Y = R$ (the set of all real numbers) then R^n is just the *n-dimensional Euclidean space.* If X is a metric space, so is X^N.

If each X is a Hausdorff space so is $X = \prod_{\alpha} X_\alpha$. The *Theorem of Tychonoff* says that X is compact if each X_α is compact. If each X_α is a uniform space, then one can define a uniform structure on X as well in such a way that the topology defined by the uniformity coincides with the product topology ([18], p. 182). The product uniform space X is complete if and only if each X_α is complete.

8. Mappings

Let X_u and Y_v be two topological spaces. A mapping f of X_u into Y_v is said to be *continuous* if for each v-open set P in Y_v,

$$f^{-1}(P) = \{x: f(x) \in P\}$$

is u-open in X_u.

If f is a continuous mapping of X_u into Y_v and g a continuous mapping of Y_v into Z_w, then the composition $g \circ f = h: X_u \to Z_w$ is a continuous mapping.

Let f be a mapping of X_u into Y_v, then the following statements are equivalent:

(*a*) f is continuous.

(*b*) For each closed set C in Y_v, $f^{-1}(C)$ is closed in X_u.

(*c*) Let V be an arbitrary member of a subbase of the topology v on Y. Then $f^{-1}(V)$ is open in X_u.

(*d*) For each $x \in X$ and each neighbourhood V of $f(x)$, $f^{-1}(V)$ is a neighbourhood of x in X_u.

(*e*) For each subset B of Y, $\overline{f^{-1}(B)} \subset f^{-1}(\bar{B})$.

(*f*) For each subset A of X, $f(\bar{A}) \subset \overline{f(A)}$.

For the proof of the equivalence of these statements, the reader can consult [18], p. 86.

Let A be a subspace of X_u with the relative topology. Let f be a continuous mapping of X_u into Y_v, then the restriction $(f \mid A)$ of f on A is also continuous.

A mapping f of X_u into Y_v is said to be *almost-continuous* at $x \in X$, if for each v-neighbourhood V of $f(x)$, $\overline{f^{-1}(V)}$ contains a u-neighbourhood of x. f is almost-continuous on X if it is so at each point $x \in X$.

Every continuous mapping is almost-continuous, but not conversely.

A mapping f of X_u into Y_v is said to be *open* if for each u-open set P in X_u, $f(P)$ is a v-open set of Y_v.

A mapping f of X_u into Y_v is said to be *almost-open* if, for each u-neighbourhood U of $x \in X_u$, $\overline{f(U)}$ contains a v-neighbourhood of $f(x) \in Y_v$.

Clearly every open mapping is almost-open. The converse is not true.

Let f be a mapping of X_u into Y_v. The subset $\{(x,f(x)): x \in E\}$ of $X \times Y$ is called the *graph* of f.

If each point in Y_v is the image of some point in X_u under the mapping f, then f is said to be *onto*. If for $x \neq y$ $(x, y \in X)$, $f(x) \neq f(y)$, then the mapping is said to be *one-to-one* (1:1).

A one-to-one mapping f of X_u onto Y_v has an inverse mapping $f^{-1}: Y_v \to X_u$. If f is one-to-one and onto, then f is continuous if and only if f^{-1} is open.

A continuous, 1:1 and open mapping of X_u onto Y_v is called a *homeomorphism* or a *topological* mapping. If there exists a homeomorphism of X_u onto Y_v, then X_u and Y_v are said to be *homeomorphic* and denoted by $X_u \cong Y_v$.

The projection mapping p_α of a topological product $X = \prod_\alpha X_\alpha$ onto X_α is continuous and open.

9. Vector spaces

Let K be a given field. A set E is said to be a *vector space* over the field K, if the following conditions are satisfied:

(*a*) E is an additive abelian group.
(*b*) For each $\lambda \in K$ and $x \in E$, $\lambda x \in E$.

Moreover, the operation defined in (*b*) satisfies the following conditions:

$$(b_1) \quad \lambda(x+y) = \lambda x + \lambda y,$$
$$(b_2) \quad (\lambda+\mu)x = \lambda x + \mu x,$$
$$(b_3) \quad \lambda(\mu x) = (\lambda\mu)x,$$
$$(b_4) \quad 1x = x,$$
$$(b_5) \quad \text{o}x = 0,$$

where $x, y \in E$; $\lambda, \mu \in K$; o is the zero of K and 0 is the zero of E.

If K is the field of the real (complex) numbers, the vector space E is said to be a real (complex) vector space.

A non-empty subset F of a vector space E is said to be a *vector subspace* of E, if F is a vector space over the same field and closed under the operations induced from E.

Let F be a subspace of a vector space E. The translate of F, i.e. $F+x$, where $x \in E$, is called a *linear manifold*.

A subset H of a real vector space E is said to be a *hyperplane* if there exists $x \in E$, $x \notin H$ such that $E = H+Rx$, where R is the set of real numbers.

If E is a vector space over the field K and $x_1, x_2,..., x_n$ are elements in E, the object $\sum_{i=1}^{n} \lambda_i x_i$, where $\lambda_i \in K$, is called a *linear combination* of $x_1,..., x_n$.

A subset F of a vector space E is called *linearly independent* if F is not empty or $\{0\}$ and no element of F is a linear combination of any finite subset of other elements in F.

A maximal linearly independent subset of a vector space is called a *Hamel basis* (or *vector basis*).

Every vector space has a Hamel basis. Any two Hamel bases of a vector space have the same cardinal number. The cardinal number of a Hamel basis of a vector space is called its *dimension*.

Let E be a vector space and F a vector subspace of E. Consider the set $G = \{x+F;\, x \in E \ni x+F = y+F \text{ if and only if } x-y \in F\}$. Define the addition operation as $(x+F)+(y+F) = (x+y)+F$ and the scalar multiplication as $\lambda(x+F) = \lambda x+F$. Then it is easy to see that G is a vector space over the same field. It is called a *quotient vector space* and denoted by E/F.

An arbitrary *product* $E = \prod_{\alpha} E_\alpha$ of vector spaces E_α is a vector space, where addition and scalar multiplication are defined as coordinatewise addition and scalar multiplication. More precisely, for each $x = (x_\alpha)$, $y = (y_\alpha)$, $x_\alpha, y_\alpha \in E_\alpha$, $x+y = (x_\alpha+y_\alpha)$ and $\lambda x = (\lambda x_\alpha)$.

Let E_α $(\alpha \in A)$ be a family of vector spaces. Let $F = \sum_{\alpha \in A} E_\alpha$ denote the set of $x = (x_\alpha)$, where $x_\alpha = 0$ for all α except for a finite subset of A. With the operation of coordinatewise addition and scalar multiplication as defined in the above paragraph, it is easy to see that F is a vector space. F is called the *direct sum* of E_α's.

A mapping f of a vector space E (over the field K) into another vector space F (over K) is said to be *linear* if for each λ, $\mu \in K$ and $x, y \in E$, $f(\lambda x+\mu y) = \lambda f(x)+\mu f(y)$.

A linear mapping of a vector space E into the field K is called a linear *functional*.

A set H in a real vector space E is a hyperplane if and only if there is a non-zero linear functional f on E such that

$$H = \{x: f(x) = \alpha\}$$

for some real number α. For a given non-zero linear functional f, the set $\{x: f(x) \leqslant \alpha\}$ is called a *half-space*.

If E is a vector space over the field R of real numbers, the set E^* of all linear functionals of E is called the *algebraic dual* of E. E^* is a vector space. If $f \in E^*$, the value of f at $x \in E$ will be denoted by $\langle x, f \rangle$.

Let E be a real vector space. Let $x, y \in E$. The set L of elements $\lambda x + (1 - \lambda) y$, $0 < \lambda < 1$ is called an *open segment* of E. If $0 \leqslant \lambda \leqslant 1$, then L is a *closed segment* of E.

2

TOPOLOGICAL VECTOR SPACES

1. Definition of a topological vector space

Definition 1. (*a*) A set E_u is said to be a *topological vector space* (or, in short, a *TVS*) over a given field K, if E_u as a pointset is a topological space and a vector space over K such that the mappings:

$$(x, y) \to x+y,$$

$$(\lambda, x) \to \lambda x$$

are continuous in both variables together for x, $y \in E$ and $\lambda \in K$.

(*b*) A TVS is said to be Hausdorff if the topology u of E_u is Hausdorff. A TVS is called a real (or complex) topological vector space according as K is the field of real (or complex) numbers.

We shall usually deal with the real Hausdorff TVS in the sequel. Most of the results can easily be proved for complex Hausdorff TVS as well. We shall use the term 'TVS' instead of 'real Hausdorff TVS'.

Some of the most useful examples of TVS are as follows:

1. $E_u = R$, the real line over the field of the reals. The topology u is generated by the set of open intervals $a < x < b$, where a and b are rational numbers.

2. $E_u = R^n$, the set of all n-tuples $(x_1, x_2, \ldots, x_n)$, where x_i is real for all $i \leqslant n$. The topology u is generated by the set of open balls:

$$B_r(x^0) = \{x = (x_1, \ldots, x_n) \colon (x_1 - x_1^0)^2 + \ldots + (x_n - x_n^0)^2 < r^2\},$$

where r is a real number. E_u is also called the Euclidean n-space. E_u is a complete metric space.

3. Let A be an arbitrary set of indices. Let E_u denote the topological product $\prod_{\alpha \in A} R_\alpha$ (Chapter 1, § 9) of real lines, where R_α is a copy of R and thus $E_u = R^A$, where u is the product topology.

If $A = N$, the set of all positive integers, R^N is the space of all real sequences. It is a TVS in the product topology. Moreover, R^N is a metric complete space because R is a metric complete space (see Chapter 1, § 7).

4. Let A be an arbitrary set of indices. Let $E_u = \sum_{\alpha \in A} E_\alpha$ denote the direct sum (see Chapter 1, § 9) of locally convex TVS's (§ 4, Def. 5 (a)) E_α with the direct sum topology u defined in (3), § 5.

If each $E_\alpha = R$, then E_u will be denoted by $R^{(A)}$. Further, if $A = N$, then $R^{(N)}$ is the space of all finite real sequences with the direct sum topology.

5. Let $E_u = L_p(N)$ $(p \geqslant 1)$ denote the vector space of all sequences $x = (x_n)$ such that $\sum_{n=1}^{\infty} |x_n|^p < \infty$. Let the topology u be the metric topology defined as follows: For each pair $x, y \in E$ let

$$d(x, y) = \sqrt[p]{\left(\sum_{i=1}^{\infty} |x_i - y_i|^p\right)}.$$

Then E_u is a complete metric TVS.

If $p = \infty$, then $E = L_\infty(N)$ is the vector space of all sequences $x = (x_n)$ such that $\sup_n |x_n| < \infty$. The metric defined by

$$d(x, y) = \sup_n |x_n - y_n|$$

makes $L_\infty(N)$ a metric TVS.

6.† Let I denote the closed unit interval $[0, 1]$. $L_p(I)$ denotes the vector space of all measurable functions f defined on $[0, 1]$ such that $\int_0^1 |f(x)|^p\, dx < \infty$. With the metric defined by the distance

$$d(f, g) = \left\{ \int_0^1 |f(x) - g(x)|^p\, dx \right\}^{1/p} \quad (p \geqslant 1),$$

$L_p(I)$ is a metric TVS.

If $p = \infty$, $L_\infty(I)$ denotes the space of all measurable functions f such that

$$\sup_{0 \leqslant x \leqslant 1} |f(x)| < \infty.$$

7.† The space of all measurable functions defined on a closed interval $[0, 1]$ is a topological vector space with the metric topology defined by the distance as follows: For each pair of measurable functions f and g,

$$d(f, g) = \int_0^1 \frac{|f(x) - g(x)|}{1 + |f(x) - g(x)|}\, dx.$$

8. Let T be a compact Hausdorff topological space. Let $C(T)$ denote the vector space of all continuous real valued functions on T. With

† In examples 6 and 7 it is understood that f, g represent the equivalence classes, i.e. any two functions f_1 and f_2 are in the same equivalence class if $f_1(x) \neq f_2(x)$ on a set of measure zero.

the topology defined by the metric:

$$d(f,g) = \sup_{x \in T} |f(x)-g(x)|,$$

$C(T)$ is a metric TVS.

(1) Let E_u be a TVS. (*a*) For a fixed element $a \in E$ and $0 \neq \lambda \in R$, the mapping $f(x) = \lambda x+a$ is a homeomorphism. Therefore a neighbourhood system of any point $a \in E$ consists of the translates of the neighbourhood system of 0. (*b*) Every TVS has a uniform structure. Hence every TVS E_u has a unique (up to an isomorphism) completion $\hat{E}_u$. (*c*) A TVS E_u is complete if and only if $E_u = \hat{E}_u$. (*d*) Every TVS is completely regular. (*e*) A subset P of E_u is precompact if and only if for each neighbourhood U of 0 there exists a finite subset G such that $P \subset G+U = \{x+y : x \in G, y \in U\}$. (*f*) If A is open, so are $A \pm B$ for any set B. However, $A \pm B$ need not be closed, even though A and B both are closed.

Let E be a TVS and F a closed subspace of E. Let E/F denote the vector space with the topology defined as follows: Let ϕ: $E \to E/F$ denote the canonical mapping, i.e. $\phi(x) = x+F$. A set P is open in E/F if and only if $\phi^{-1}(P)$ is open in E. This way we get a topology on E/F, and E/F in this topology is called a *quotient* TVS.

(2) The canonical mapping ϕ of E onto E/F is linear, continuous, and open.

2. Neighbourhood systems in a TVS

Definition 2. Let E be a vector space over the field R.

(*a*) A subset M of E is said to be *circled* if for each $x \in M$, $\lambda x \in M$ for all $|\lambda| \leqslant 1$, $\lambda \in R$.

(*b*) A subset A of E is said to *absorb* another subset B of E, if there exists $\alpha > 0$ such that $\lambda B \subset A$ for all $|\lambda| \leqslant \alpha$, $\lambda \neq 0$.

(*c*) A subset B of E is said to be *absorbing* if it absorbs each point of E.

(*d*) Let M be a non-empty subset of E. The smallest circled set containing M is called the *circled hull* of M.

THEOREM 1. *In a TVS E_u there exists a fundamental system $\mathcal{U}$ of u-closed neighbourhoods of the origin* 0 *such that:*

(1°) *Each U in $\mathcal{U}$ is circled and absorbing.*

(2°) *For each U in $\mathcal{U}$ there exists a V in $\mathcal{U}$ such that $V+V \subset U$.*

(3°) $\bigcap U = \{0\}$, *where U is in $\mathcal{U}$.*

Conversely, if E is a real vector space and $\mathcal{U}$ a filter-base satisfying conditions (1°) *to* (3°), *then there exists a topology u on E such that E_u is a TVS and $\mathcal{U}$ is a fundamental system of neighbourhoods of* 0.

Proof. (*a*) Let $\mathscr{V} = \{V\}$ be a fundamental system of u-open neighbourhoods of 0. Since E_u is regular, one can replace $\mathscr{V}$ by $\mathscr{U}$ consisting of u-closed neighbourhoods U of 0. Since each open set containing 0 is absorbing, and since it is easily seen that the circled hull of a non-empty set is circled, each U can be taken to satisfy (1°). Since the mapping: $(x, y) \to x+y$ is continuous at $(0, 0)$, $\mathscr{U}$ satisfies (2°). Finally E_u being Hausdorff, condition (3°) is satisfied.

(*b*) Now let E be a vector space and $\mathscr{U}$ a filterbase satisfying (1°) to (3°). Since each U in $\mathscr{U}$ is circled, $U = -U$ and therefore $0 \in U$. Now (1°), coupled with (2°), shows that $\mathscr{U}$ is a fundamental system of neighbourhoods of 0 for a topology u such that the mapping $(x, y) \to x+y$ is continuous ([2], Chapitre III, § 1, N° 2). In other words, for each U there exists a $V \ni V+V \subset U$. From this it follows that $2V_1 \subset U$ for some V_1 in $\mathscr{U}$ and a given U. Inductively proceeding we obtain $2^n V_n \subset U$ $(n > 0)$. Now let n be so chosen that $|\lambda 2^{-n}| \leqslant 1$ for any given $\lambda \neq 0$. Then, by putting $V_n = W$ and observing that U is circled we get $\lambda W \subset U$ which easily leads to the continuity of the mapping $(\lambda, x) \to \lambda x$. Condition (3°) evidently implies that the topology u is Hausdorff. Thus the proof is completed.

Definition 3. (*a*) A TVS E_u is said to be *metrizable* if there exists a countable fundamental system of neighbourhoods of 0.

These neighbourhoods can be chosen to satisfy conditions (1°) to (3°) of Theorem 1.

(*b*) A metrizable complete TVS is called an *F-space*.

3. The Hahn-Banach theorem

Definition 4. (*a*) Let E be a vector space over R. A non-empty set M is said to be *convex* if for each pair $x, y \in M$, $\lambda x+(1-\lambda)y \in M$, $0 \leqslant \lambda \leqslant 1$.

The intersection of a family of convex sets in E is either empty or a convex set.

(*b*) Let B be any non-empty subset of E. The intersection of all convex sets containing B is called the *convex hull* of B.

(*c*) Let B be any non-empty subset of a TVS E_u. The intersection of all u-closed and convex sets containing B is called the *convex closure* of B.

(*d*) A subset A of E is said to be a *cone with* (*or without*) *vertex* 0 if for all $\lambda \geqslant 0$ $(\lambda > 0)$, $\lambda A \subset A$.

(1) (*a*) Each linear manifold and each hyperplane of E is a convex set. (*b*) Every subspace of E is also convex.

(2) The convex closure of a set B in E, being the convex hull of E, is convex and, being the intersection of closed sets, is closed.

(3) (*a*) If A is a cone with vertex 0, then $\bar{A}$ is also a cone with vertex 0. (*b*) A subset A of E is a convex cone with vertex 0 if and only if $A+A \subset A$ and for all $\lambda \geqslant 0$, $\lambda A \subset A$.

LEMMA 1. *Let A be a non-empty open convex set in R^2 such that $0 \notin A$. Then there exists a straight line H (in R^2) which passes through 0 and does not meet A.*

Proof. Let $B = \bigcup_{\lambda>0} \lambda A$. Being the union of open sets, B is open. It is easy to see that B is a convex cone and does not contain 0. In order to prove the lemma, it is sufficient to show that there exists $x \in R^2$ such that $x \notin B$ and $-x \notin B$, because then the straight line passing through x and $-x$ will be the required straight line. Clearly $B \neq R^2 \setminus \{0\} = G$, because otherwise, due to convexity of B, $0 \in B$ which is a contradiction. Let x be a boundary point of B (in G) which is non-empty. $x \notin B$ (because B is open in G) and $-x \notin B$ because otherwise the open segment joining x and $-x$ would contain 0 and therefore 0 would belong to B which is contrary to the assumption.

Observe that Lemma 1 is clearly true in R^n $(n \geqslant 2)$.

THEOREM 2 (*Hahn–Banach*). *Let E be a TVS, A a non-empty open convex subset of E, and M a linear manifold which does not meet A. Then there exists a closed hyperplane H which contains M and does not meet A.*

Proof. Without loss of generality it can be assumed that M is a subspace of E. Let $\mathscr{H}$ denote the set of all vector subspaces which contain M and do not meet A. $\mathscr{H}$ contains M. Let us introduce partial-ordering by inclusion. By Zorn's lemma, there exists a maximal subspace H. For every vector subspace K in $\mathscr{H}$, clearly $\bar{K}$ is in $\mathscr{H}$. Since H is maximal, $\bar{H} = H$. Now in order to show that H is a hyperplane, we have to show that E/H has dimension 1. Let $\varphi\colon E \to E/H$ be the canonical mapping. $\varphi(A)$ is a non-empty convex open set of E/H and does not contain 0. Suppose E/H has dimension at least equal to 2. Then by Lemma 1, there exists a straight line L in E/H, passing through 0 and such that $L \cap \varphi(A) = \emptyset$. Now consider

$$\varphi^{-1}(L) = L+H \supset H.$$

Since $\varphi^{-1}(L) \cap A = \emptyset$ and H is maximal, therefore a contradiction establishes the theorem.

(4) Theorem 2 can be used to show that in a TVS two disjoint sets A and B, where A is a non-empty convex open set and B any non-empty convex set, can be separated by a closed hyperplane.

4. Locally convex topological vector spaces

Definition 5. (*a*) A TVS is said to be *locally convex* (or, in short, an *l.c. space*) if there exists a fundamental system of convex neighbourhoods of 0.

(*b*) An l.c. space which is metrizable and complete will be called a *Fréchet space*.

(1) (*a*) There exists a metrizable TVS which is not locally convex, namely Example 7 (§ 1). (*b*) There exists an l.c. space which is not metrizable, e.g. R^A of Example 3 (§ 1), where A is an uncountable set of indices.

Definition 6. (*a*) Let E be a real vector space. A real valued function p defined on E is said to be a *semi-norm* if the following conditions are satisfied:

(1°) $p(\lambda x) = |\lambda| p(x)$, for all $\lambda \in R$ and $x \in E$.

(2°) $p(x+y) \leqslant p(x)+p(y)$.

Clearly $p(0) = 0$ and $p(x) \geqslant 0$ for each x.

(*b*) A semi-norm is said to be a *norm* provided $p(x) = 0$ if and only if $x = 0$.

(*c*) A norm is denoted by $\|...\|$ and satisfies (1°) and (2°).

A norm is always a semi-norm, but not conversely.

(2) Let E_u be a TVS. There is a 1:1 correspondence between the set of all convex circled closed subsets (of E) containing 0 as their interior point, and the set of all continuous semi-norms p defined on E_u.

Since there exists a fundamental system of convex circled closed neighbourhoods of 0 in an l.c. space E_u, the topology u can also be defined by a subset of all continuous semi-norms on E_u ([3], Chapitre II, § 4).

(3) An l.c. space E_u is metrizable if and only if u can be described by a countable set of continuous semi-norms.

(4) (*a*) The set of all semi-norms on a vector space E defines a topology which is the *finest locally convex* topology denoted by ω. (*b*) In an l.c. space each convex closed set A is the intersection of closed half-spaces containing A. (*c*) In an l.c. space each closed linear manifold M is the intersection of closed hyperplanes containing M.

Definition 7. (*a*) A TVS E_u is a *normed space* if u can be defined by a norm.

(*b*) A TVS is a *Banach space* if it is normed and complete in the metric induced by the norm.

(5) (*a*) A norm on a TVS induces a metric and therefore a uniform structure. (*b*) A normed space (in particular a Banach space) is locally

convex. (*c*) Every l.c. space is a subspace of a product of Banach spaces. (*d*) Every complete l.c. space is a closed subspace of a product of Banach spaces.

Proposition 1. *Let* A_i $(1 \leqslant i \leqslant n)$ *be a finite family of compact subsets of a TVS* E. *Then the convex hull of* $A = \bigcup_{i \leqslant n} A_i$ *is also compact.*

Proof. By Tychonoff's theorem, $\prod_{i \leqslant n} A_i$ is compact and the convex hull of A is the set of all $x = \sum_{i \leqslant n} \lambda_i x_i$ where $\lambda_i \geqslant 0$, $\sum_{i \leqslant n} \lambda_i = 1$, and $x_i \in A_i$. That means the convex hull of A is the image of the set $B \times \prod_{i \leqslant n} A_i$, where B is a subset of R^n defined by $\lambda_i \geqslant 0$, $\sum_{i \leqslant n} \lambda_i = 1$. B is clearly compact, and the mapping:

$$(\lambda_1,\ldots,\lambda_n, x_1,\ldots,x_n) \to \sum \lambda_i x_i$$

is easily seen to be continuous. Therefore the convex hull of A is compact.

Proposition 2. *Let* E *be an l.c. space and* A *a precompact subset of* E. *Then the convex circled hull of* A *is also precompact.*

Proof. It is easy to see that the circled hull of A is precompact. Since the convex circled hull of a set is the convex hull of its circled hull, we show that the convex hull of A is precompact. For any convex neighbourhood V of 0 in E, there exists a finite subset a_i $(1 \leqslant i \leqslant n)$ of E such that $A \subset \bigcup_{i \leqslant n} (a_i + V) = C$. The convex hull of A is, therefore, contained in the convex hull of C. But the convex hull of C is equal to $B+V$, where B is the convex hull of a_i $(1 \leqslant i \leqslant n)$. By Proposition 1, B is compact and therefore there exists a finite subset x_k $(1 \leqslant k \leqslant m)$ such that $B \subset \bigcup_{k \leqslant m} (x_k + V)$. Therefore $B + V \subset \bigcup_{k \leqslant m} (x_k + 2V)$ which proves that the convex hull of A is precompact.

The following corollary is immediate from the above proposition.

Corollary 1. *In a complete l.c. space, the convex closure of a compact set is compact.*

The Hahn–Banach theorem in its often-used analytic form is stated as follows:

Theorem 3. *Let* E *be a real vector space and* p *a semi-norm on* E. *Let* M *be a vector subspace of* E *and* f *a linear functional defined on* M *such that* $|f(x)| \leqslant p(x)$ *for all* $x \in M$. *Then there exists a linear functional* $\tilde{f}$ *defined on* E *which coincides with* f *on* M *and* $|\tilde{f}(x)| \leqslant p(x)$ *for all* $x \in E$.

The proof of this theorem is omitted. It immediately follows from Theorem 2 if one defines a topology (on E) in which p is continuous, and observes that the linear functional f on M gives a linear manifold of E, and the set $\{x: p(x) < 1\}$ is a non-empty convex open subset of E.

5. Inductive limits of l.c. spaces

Definition 8. (*a*) Let E_α ($\alpha \in A$) be a family of l.c. spaces and f_α a linear mapping of E_α into a vector space E, for each α. Suppose $E = \bigcup_{\alpha \in A} f_\alpha(E_\alpha)$. Let u be the *finest* locally convex topology on E such that f_α is continuous for each α. The vector space E, endowed with the topology u (in short, E_u), is called the *inductive limit* of E_α.

(1) A fundamental system of neighbourhoods U of 0 in E_u is described as follows: A circled, convex, and absorbing subset U of E is a u-neighbourhood of 0 if and only if, for each α, $f_\alpha^{-1}(U)$ is a neighbourhood of 0 in E_α under the given topology on E_α.

(*b*) If, in Definition 8 (*a*), the set $A = N$ (the set of positive integers), then E_u is called the *generalized strict inductive limit* of E_n.

In addition, if each f_n is the identity mapping or, in other words, E is the union of a strictly increasing sequence of its subspaces E_n ($n \geqslant 1$), then the generalized strict inductive limit topology u of E induces a topology coarser than the initial topology of E_n for each n.

(*c*) If, in (*b*) where f_n is the identity mapping, the inductive limit topology on E_u induces the same topology as that of E_n, then E_u is called the *strict inductive limit* of E_n.

(2) (*a*) A fundamental system of neighbourhoods of 0 in a strict inductive limit E of E_n ($n \geqslant 1$) is described as follows: A convex, circled, and absorbing set U in E is a u-neighbourhood of 0 in E if and only if for each n, $U \cap E_n$ is a neighbourhood of 0 in the given topology of E_n ($n \geqslant 1$). (*b*) If E is the strict inductive limit of l.c. spaces E_n, then the strict inductive limit is Hausdorff if each E_n is Hausdorff.

(*d*) A (generalized) strict inductive limit of Fréchet spaces F_n ($n \in N$) is called a (*generalized*) *LF-space*.

(3) The direct sum $E = \sum E_\alpha$ of a family E_α ($\alpha \in A$) of l.c. spaces, endowed with the finest l.c. topology u such that the embedding $E_\alpha \to E$ is continuous for each α, is the inductive limit of E_α. u is called the *direct sum* topology.

PROPOSITION 3. *Let f be a linear mapping of an inductive limit E_u of E_α ($\alpha \in A$) into an l.c. space F. Then f is continuous if and only if the composition mapping $f \circ f_\alpha$ is continuous for each α.*

Proof. If f is continuous, then $f \circ f_\alpha$ is obviously continuous (because f_α is continuous by definition). On the other hand, for any convex circled neighbourhood V of 0 in F, $f_\alpha^{-1}(f^{-1}(V))$ being a neighbourhood (because $f \circ f_\alpha$ is continuous) of 0 in E_α implies $f^{-1}(V)$ is a neighbourhood of 0 in E due to the definition of inductive limits. Therefore f is continuous.

PROPOSITION 4. *Let E_u be the strict inductive limit of E_n $(n \geqslant 1)$. If, for each n, E_n is closed in E_{n+1}, then E_n is a closed subspace of E.*

Proof. If E_n is closed in E_{n+1}, then E_n is closed in E_{n+2} because E_{n+1} is closed in E_{n+2}. Repeating the argument a finite number of times we see that E_n is closed in E_{n+m} $(m \geqslant 1)$. Clearly $\bigcup_{m \geqslant 1} E_{n+m} = E$. Let $x \notin E_n$. Then there exists a neighbourhood V_{n+m} of 0 in E_{n+m} such that $(x+V_{n+m}) \cap E_n = \emptyset$. But then there exists a neighbourhood V of 0 in E such that $V \cap E_{n+m} = V_{n+m}$. Now $(x+V) \cap E_{n+m} = x+V_{n+m}$, so $(x+V) \cap E_n = \emptyset$ proves the proposition.

(4) (*a*) Let E be the strict inductive limit of E_n $(n \geqslant 1)$. If each E_n is complete, so is E ([3], Chap. II, § 2, N° 5, Exer. 9). (*b*) Each LF-space is complete.

6. Barrelled and bornological spaces

Definition 9. (*a*) Let E_u be an l.c. space. A subset B of E_u which is closed, convex, circled, and absorbing is called a *barrel*.

In an l.c. space each convex, closed, and circled neighbourhood of 0 is a barrel. The converse is not true in general.

(*b*) An l.c. space E_u is called a *barrelled space* or a *t-space* (in French: *espace tonnelé*) if every barrel in E_u is a neighbourhood of 0.

(1) (*a*) For any barrel B in an l.c. space E_u, $\bigcup_{n>0} nB = E$, because B is absorbing. If E_u were of the second category in itself, or in other words, a *Baire space*, then, for some n_0, $n_0 B$ has to have an interior point. From this and using convexity of B, it follows that B is a neighbourhood of 0. In other words, E_u is a t-space. (*b*) Since a Fréchet space (in particular a Banach space) is of the second category by Baire's theorem (see Chapter 1), therefore, by the above argument, each Fréchet space is a t-space.

(2) (*a*) Inductive limits of t-spaces are t-spaces. (*b*) Each quotient space of a t-space is also a t-space. (*c*) Any topological direct sum of t-spaces is a t-space. (*d*) However, a closed subspace of a t-space need *not* be a t-space. (*e*) Let E_u be any l.c. space. E with the finest locally convex topology ω (see § 4) is a t-space ([4], Chapitre III, § 2).

Barrelled spaces are most suited for carrying over many important theorems (e.g. Banach–Steinhaus, open mapping and closed graph theorems) of Functional Analysis. They form a natural generalization of Fréchet spaces. A barrelled space is neither necessarily metrizable nor complete ([4], Chapitre III, § 1, Ex. 3 and 6).

PROPOSITION 5. *Let E_u and F_v be two l.c. spaces. Let f be a linear, continuous and almost open mapping of E_u into F_v. If E_u is barrelled, so is F_v.*

Proof. Let B be a v-barrel in F_v. Since f is linear, $f^{-1}(B)$ is convex, circled, and absorbing. Further, f being continuous implies $f^{-1}(B)$ is also closed. In other words, $f^{-1}(B)$ is a u-barrel and therefore a u-neighbourhood of 0, because E_u is a barrelled space by hypothesis. Now f being almost open implies

$$\overline{f(f^{-1}(B))} = \bar{B} = B$$

is a v-neighbourhood of 0. This proves the proposition.

PROPOSITION 6. *Let E_u be a barrelled space and F_v any l.c. space. Then:*

(*a*) *Any linear mapping f of E_u into F_v is almost continuous.*

(*b*) *Any linear mapping g of F_v onto E_u is almost open.*

Proof. (*a*) Let V be a convex, closed, and circled v-neighbourhood of 0 in F_v. $\overline{f^{-1}(V)}$ is convex, circled, and closed due to linearity of f and the fact that the convexity and circledness properties are preserved under the closure-operation. Moreover, $\overline{f^{-1}(V)}$ is absorbing because V is absorbing in F_v and f is linear. Therefore $\overline{f^{-1}(V)}$, being a barrel, is a neighbourhood of 0 in E_u because E_u is a t-space. Hence f is almost continuous.

(*b*) Let V be a closed, circled, and convex neighbourhood of 0 in F_v. $\overline{g(V)}$ is a barrel for the same reasons as in (*a*) plus the assumption that g is onto. Therefore $\overline{g(V)}$ is a u-neighbourhood of 0, because E_u is barrelled. This shows that g is almost open.

Definition 10. (*a*) A subset B of an l.c. space E_u is said to be *bounded* if B is absorbed by each neighbourhood of 0.

(*b*) Let $\mathscr{B}$ be a subfamily of all bounded sets of an l.c. space E_u. $\mathscr{B}$ is said to be a *fundamental system of bounded sets* if for each bounded set B in E_u, there exists a set A in $\mathscr{B}$ such that $B \subset A$.

(3) (*a*) In an l.c. space E_u, the circled hull and convex closure of a bounded set are bounded. (*b*) Each precompact set in E_u is bounded. (*c*) In particular, each Cauchy sequence in E_u is bounded. (*d*) Each point of E_u is bounded. For proofs see ([4], Chapitre III, § 2).

Proposition 7. *In a TVS E, for a set B in E to be bounded it is necessary and sufficient that for each sequence x_n $(n \geqslant 1)$ in B, and each sequence of numbers λ_n $(n \geqslant 1)$, $\lambda_n \geqslant 0$, $\lambda_n \to 0$ implies $\lambda_n x_n \to 0$.*

Proof. Since B is bounded, for each sequence x_n $(n \geqslant 1)$ there exists $\alpha > 0$ such that for all $|\lambda| \leqslant \alpha$, $\lambda \neq 0$, $\lambda x_n \in V$, for any arbitrary circled neighbourhood V of 0 and for each n. Now there exists n_0 such that for all $n \geqslant n_0$, $\lambda_n \leqslant \alpha$. Since V is circled and arbitrary, $\lambda_n x_n \in V$ for $n \geqslant n_0$ implies that $\lambda_n x_n \to 0$.

To show that the condition is sufficient, suppose B is not bounded. That means there exists a circled neighbourhood U of 0 in E such that $B \not\subset nU$ for all $n \geqslant 1$. Thus for each n there exists $x_n \in B$ such that $x_n \notin nU$; or $(1/n)x_n \notin U$. Hence $(1/n)x_n \not\to 0$. Therefore if the condition of the proposition holds then B must be bounded.

Proposition 8. *Let E be the strict inductive limit of an increasing sequence of l.c. spaces E_n $(n \geqslant 1)$ (see § 5). A subset B in E is bounded if and only if B is in E_m, for some m, and B is bounded in E_m.*

Proof. Since the strict inductive limit topology on E induces the given topology on E_n, for each n, a bounded set B in E_n is clearly bounded in E. For the converse, let B be a bounded set of E and suppose that there is a sequence $x_n \in B$ such that $\{x_n\} \not\subset E_m$ for any m. In view of Proposition 7, we wish to show that $x_n \not\to 0$. For each m there exists $x_{n(m)} \notin E_m$ but $x_{n(m)} \in E_{m+1}$. One can suppose that the sequence $n(m)$ is strictly increasing. Since one can find a neighbourhood V_m of 0 in E_m for each m such that $V_{m+1} \cap E_m = V_m$, $x_{n(m)} \notin V_m$. Since $\bigcup_{m \geqslant 1} V_m = V$ is a neighbourhood of 0 in E, the subsequence $(x_{n(m)})$ does not belong to V. In other words, $x_{n(m)} \not\to 0$ as $n(m) \to \infty$. This establishes the proposition.

In general for inductive limits E of E_α ($\alpha \in A$, A is an arbitrary set of indices) the above proposition is not true.

Corollary 2. *Let E be the strict inductive limit of a strictly increasing sequence of l.c. spaces E_n $(n \geqslant 1)$. If each E_n has a countable fundamental system of bounded sets, so does E.*

Proof. The corollary is immediate from the above proposition.

Definition 11. (*a*) A TVS E_u is said to be *quasi-complete* if every closed bounded set of E_u is complete.

(*b*) Let E_u be an l.c. space. A subset M of E_u is said to be *bornivorous*† if M absorbs every bounded set of E_u.

† The term has been introduced by the authors of [25].

(*c*) An l.c. space is said to be *bornological* if each convex set which is bornivorous is a neighbourhood of 0.

(*d*) An l.c. space E_u is said to be *quasi-barrelled* if each barrel of E_u which is bornivorous is a neighbourhood of 0.

(4) (*a*) Each complete TVS is quasi-complete. (*b*) A metrizable quasi-complete TVS is an F-space. (*c*) In a quasi-complete l.c. space, the convex closure of each precompact set is compact. (*d*) Each closed subspace of a quasi-complete l.c. space is quasi-complete. (*e*) The strict inductive limit of quasi-complete l.c. spaces is quasi-complete. (*f*) An arbitrary product of quasi-complete (complete) l.c. spaces is quasi-complete (complete) ([4], Chapitre III, § 2). (*g*) However, a quotient space of a quasi-complete (complete) l.c. space need not be quasi-complete (complete) [21].

(5) (*a*) The inductive limits of bornological (quasi-barrelled) spaces are bornological (quasi-barrelled). (*b*) In particular, quotient spaces and direct sums are also of the same type. (*c*) A countable product of bornological (quasi-barrelled) spaces is bornological (quasi-barrelled) ([4], Chapitre III, § 2). (*d*) Proposition 5 remains true for quasi-barrelled spaces, as is easy to see.

PROPOSITION 9. *Each barrelled and/or bornological space is quasi-barrelled. But the converse is not true.*

Proof. The first statement follows trivially from the definition. For the converse, let E be a barrelled space which is not bornological ([26], [33]) and F a bornological space which is not barrelled (such a space exists, e.g. $R^{(N)}$ with the norm topology). Then $E \times F$ is quasi-barrelled ((5) (*c*)). However, if B is a bornivorous convex set in E which is not a neighbourhood of 0, then $B \times F$ is a bornivorous convex subset of $E \times F$ which is not a neighbourhood of 0. In other words, $E \times F$ is not bornological. Similarly if B_1 is a barrel in F which is not a neighbourhood of 0, then $E \times B_1$ will be a barrel of $E \times F$ but not a neighbourhood of 0. In other words, $E \times F$ is not barrelled.

This example is due to Mahowald and Gould [25].

7. $\mathfrak{S}$-topologies and the principle of uniform boundedness

Let E and F be two topological vector spaces. The set of all linear and continuous mappings of E into F forms a vector space and is denoted by $L(E, F)$.

Let $\mathfrak{S}$ be a class of subsets of E. One can define a topology in $L(E, F)$ of uniform convergence over sets M in $\mathfrak{S}$ as follows: Let V be a neighbourhood of 0 in F and M in $\mathfrak{S}$. Let $T(M, V)$ denote the set of all linear and continuous mappings f in $L(E, F)$ such that $f(M) \subset V$. The collection of $T(M, V)$, when M runs over $\mathfrak{S}$ and V over neighbourhoods of 0 in F, forms a sub-basis for a topology called the $\mathfrak{S}$-*topology* ([4], Chapitre III, § 3). The vector space $L(E, F)$ with an $\mathfrak{S}$-topology is denoted by $L_{\mathfrak{S}}(E, F)$.

(1) (*a*) $L(E, F)$, endowed with an $\mathfrak{S}$-topology, is a topological vector space if and only if $f(M)$ is bounded in F for each M in $\mathfrak{S}$ and $f \in L(E, F)$. (*b*) In addition, if F is an l.c. space, the $\mathfrak{S}$-topology is also locally convex and then $L(E, F)$ is an l.c. space. (*c*) If $\mathfrak{S}$ consists of bounded sets of E such that $\bigcup M$ is total in E, and F is an l.c. space then the $\mathfrak{S}$-topology is a Hausdorff locally convex topology.

The most important particular cases of the $\mathfrak{S}$-topology are as follows:

(1°) $\mathfrak{S}$ consists of all finite subsets of E. In this case, the $\mathfrak{S}$-topology is called the topology of *simple convergence*.

(2°) $\mathfrak{S}$ consists of all compact subsets of E. The $\mathfrak{S}$-topology is called the topology of *compact convergence* or the *uniform convergence topology* over compact sets.

(3°) $\mathfrak{S}$ consists of all precompact subsets of E. The $\mathfrak{S}$-topology is called the topology of *precompact convergence*.

(4°) $\mathfrak{S}$ consists of all bounded subsets of E. The $\mathfrak{S}$-topology is known as the uniform convergence topology over *bounded* sets.

If E, F are normed spaces, the topology of bounded convergence is a normed topology, the norm being defined as:

$$\|f\| = \sup_{\|x\| \leqslant 1} \|f(x)\|.$$

(2) If E and F are two l.c. spaces such that F is complete, then every linear continuous mapping f of E into F can be extended uniquely to a linear continuous mapping $\hat{f}$ of $\hat{E}$ into F. Thus $L(E, F)$ and $L(\hat{E}, F)$ are algebraically isomorphic. The $\mathfrak{S}$-topology on $L(E, F)$ can be identified with the $\mathfrak{S}$-topology on $L(\hat{E}, F)$. Further, the $\mathfrak{S}$-topology on $L(E, F)$ remains unchanged if one replaces $\mathfrak{S}$ by $\mathfrak{S}'$, where $\mathfrak{S}'$ consists of $\bar{M}$ (closure in $\hat{E}$), M in $\mathfrak{S}$.

A subset H of $L(E, F)$ is said to be $\mathfrak{S}$-*bounded* if H is bounded in the $\mathfrak{S}$-topology. In particular, H is *simply bounded* if H is bounded in the simple convergence topology of $L(E, F)$.

Proposition 10. *A set H in $L_{\mathfrak{S}}(E, F)$ is $\mathfrak{S}$-bounded if and only if, for each neighbourhood V of 0 in F, $\bigcap_{f \in H} f^{-1}(V)$ absorbs each set M in $\mathfrak{S}$.*

Proof. For each neighbourhood V of 0 in F, and for each set M in $\mathfrak{S}$, $T(M, V)$ is a neighbourhood of 0 for the $\mathfrak{S}$-topology on $L(E, F)$. Therefore, if H is $\mathfrak{S}$-bounded, there exists $\alpha > 0$ such that $\lambda H \subset T(M, V)$ for all $|\lambda| \leqslant \alpha$, $\lambda \neq 0$. Or, $\lambda f(M) \subset V$, for each $f \in H$. Hence $\lambda M \subset f^{-1}(V)$ for each $f \in H$. This shows that $\bigcap_{f \in H} f^{-1}(V)$ absorbs each M in $\mathfrak{S}$.

Conversely, for each M in $\mathfrak{S}$, there exists $\alpha > 0$ such that $\lambda f(M) \subset V$ for each $f \in H$ and for all $|\lambda| \leqslant \alpha$, $\lambda \neq 0$. Hence $\bigcup_{f \in H} \lambda f(M) \subset V$. That means $\bigcup_{f \in H} f(M)$ is bounded in F for each M in $\mathfrak{S}$, since V is an arbitrary neighbourhood of 0 in F. This shows that $\lambda H \subset T(M, V)$. In other words, H is $\mathfrak{S}$-bounded.

(3) If $\mathfrak{S}$ and $\mathfrak{S}'$ are two collections of sets in E such that $\mathfrak{S} \supset \mathfrak{S}'$, then the $\mathfrak{S}$-topology is finer than the $\mathfrak{S}'$-topology on $L(E, F)$. In other words, the identity mapping: $L_{\mathfrak{S}}(E, F) \to L_{\mathfrak{S}'}(E, F)$ is continuous. Hence every $\mathfrak{S}$-bounded set is $\mathfrak{S}'$-bounded. But the converse need not be true.

LEMMA 2. *Each barrel B of an l.c. space E_u absorbs each convex circled complete bounded subset M of E_u.*

Proof. Without loss of generality, one can assume that M generates E; otherwise one considers a subspace of E which is generated by M. In other words, M is absorbing. (Obviously M is a neighbourhood of 0 under the finest l.c. topology ω of E.) Let p be the semi-norm determined by M. Let $x \neq 0$. Since u is a Hausdorff topology, there exists a neighbourhood U of 0 such that $x \notin U$. M being bounded, there exists a $\lambda > 0$ such that $\lambda M \subset U$. Therefore $p(x) \neq 0$. This proves that p is a norm on E. Let v denote the norm topology on E defined by p. Since M is u-bounded by hypothesis, $v \supset u$.

Let x_n $(n \geqslant 1)$ be a Cauchy sequence of E_v. Then, for a given $\epsilon > 0$, there exists $n_0 = n_0(\epsilon)$ such that $p(x_{n+k} - x_n) < \epsilon$ for all $n \geqslant n_0$, $k \geqslant 1$. In other words, $x_{n+k} \in x_n + \epsilon M$ for all $n \geqslant n_0$, $k \geqslant 1$. Fixing n, x_{n+k} $(k \geqslant 1)$ is a Cauchy sequence of E_v and therefore of E_u, because $v \supset u$. Since M is complete in E_u, there exists $x \in x_n + \epsilon M$ and $x_{n+k} \to x$. Hence $p(x - x_n) < \epsilon$ for all $n \geqslant n_0$. In other words, E_v is a Banach space and therefore a barrelled space. Now clearly B is a barrel of E_v and therefore a v-neighbourhood of 0 (§ 6, Def. 9 (*b*)). M being bounded implies that B absorbs M.

THEOREM 4. (*Principle of uniform boundedness.*) *Let E and F be two l.c. spaces. Let $\mathfrak{S}$ consist of convex, circled, complete bounded subsets of E. Then every simply bounded subset H of $L_{\mathfrak{S}}(E, F)$ is $\mathfrak{S}$-bounded.*

Proof. For each convex, circled, and closed neighbourhood V of 0 in F, consider $B = \bigcap_{f \in H} f^{-1}(V)$. Clearly B is convex, circled, and closed. Since H is simply bounded, B absorbs finite subsets of E. In other words, B is a barrel in E. By Lemma 2, B absorbs each set M in $\mathfrak{S}$. Due to Proposition 10, H is $\mathfrak{S}$-bounded.

COROLLARY 3. *Let E and F be Banach spaces and H a set of linear continuous mappings of E into F. If for each $x \in E$, $\sup_{f \in H} \|f(x)\| < \infty$, then $\sup_{f \in H} \|f\| < \infty$.*

Proof. The corollary immediately follows from Theorem 4.

8. The Banach-Steinhaus theorem

Definition 12. Let E and F be two topological vector spaces. A subset H of $L(E, F)$ is said to be *equicontinuous* if, for each neighbourhood V of 0 in F, $\bigcap_{f \in H} f^{-1}(V)$ is a neighbourhood of 0 in E.

(1) Suppose E and F are l.c. spaces. (*a*) The convex circled hull of an equicontinuous set is equicontinuous ([4], Chapitre III, § 3, N° 5). (*b*) Since $L(E, F)$ is a subspace of F^E, the closure of an equicontinuous (closure taken in F^E) set is equicontinuous ([4], Chapitre III, § 3, Prop. 4). (*c*) Every equicontinuous set H of $L_{\mathfrak{S}}(E, F)$, where $\mathfrak{S}$ consists of finite subsets of E, is relatively compact ([4], Chapitre III, § 3, Cor. to Prop. 4). (*d*) Every equicontinuous set H in $L(E, F)$ is $\mathfrak{S}$-bounded.

PROPOSITION 11. *Let E and F be two l.c. spaces and H an equicontinuous subset of $L(E, F)$. Then on H the following topologies coincide:*

(*a*) *The simple convergence topology.*

(*b*) *The $\mathfrak{S}$-topology, where $\mathfrak{S}$ consists of all precompact sets of E.*

Proof. Clearly the $\mathfrak{S}$-topology induces a finer topology on H than that induced by the simple convergence topology. For the converse, let V be a convex neighbourhood of 0 in F. Then $V = \frac{1}{2}V + \frac{1}{2}V$. Since H is equicontinuous, there exists a convex circled neighbourhood U of 0 in E such that $f(U) \subset \frac{1}{2}V$ for all $f \in H$. Let A be a precompact set. Then there exists a finite set G such that $A \subset G + U$. Now if given $W' = \{f : f(A) \subset \frac{1}{2}V, f \in H\}$, a member of the basis of the $\mathfrak{S}$-topology on H, then for each $f \in W'$ and all $x \in G$,

$$f(x) \in f(A) + f(U) \subset \tfrac{1}{2}V + \tfrac{1}{2}V = V$$

completes the proof of the proposition.

COROLLARY 4. *Let H be an equicontinuous subset of $L(E, F)$. If $\mathscr{H}$ is a filter on H which converges to a mapping $f_0 \in F^E$ in the simple convergence topology, then $f_0 \in L(E, F)$ and $\mathscr{H}$ converges to f_0 uniformly on each precompact subset of E.*

Proof. By ((1) (b)), $\bar{H}$ (closure taken in F^E) is equicontinuous. Now in order to show that $f_0 \in L(E, F)$, it is sufficient to show that f_0 is linear because f_0 is continuous. Since the mapping: $F^E \to F$ is continuous, for each $\lambda, \mu \in R$ and $x, y \in E$, the set of f such that

$$f(\lambda x+\mu y)-\lambda f(x)-\mu f(y) = 0$$

is closed (because F is Hausdorff and therefore $\{0\}$ is closed). Therefore the set of all linear mappings of E into F is closed. Hence f_0 is linear. This proves the first statement. The second one follows from Proposition 11.

THEOREM 5. *Let E be a barrelled space and F any l.c. space. Then each simply bounded subset H of $L(E, F)$ is equicontinuous.*

Proof. For each convex, circled, closed neighbourhood V of 0 in F, $B = \bigcap_{f \in H} f^{-1}(V)$ is convex, circled, and closed. Since H is simply bounded, B is absorbing. In other words, B is a barrel in E and therefore a neighbourhood of 0 in E, because E is barrelled. This shows, by definition, that H is equicontinuous.

A consequence of this theorem is what is generally known as the Banach–Steinhaus Theorem.

COROLLARY 5. (*Banach–Steinhaus theorem.*) *Let E be a barrelled space and F any l.c. space. Let $\mathscr{H}$ be a simply bounded filter on $L(E, F)$ which converges simply to $f_0 \in F^E$. Then $f_0 \in L(E, F)$ and $\mathscr{H}$ converges to f_0 uniformly on compact sets.*

Proof. Since $\mathscr{H}$ is a simply bounded filter or, in other words, $\mathscr{H}$ contains a simply bounded set H, H is equicontinuous by Theorem 5, and hence the corollary follows from Corollary 4.

Corollary 5 remains true if the filter $\mathscr{H}$ has a countable basis. For then $\mathscr{H}$ is an elementary filter associated with a sequence f_n $(n \geqslant 1)$ ([2], Chapitre I, § 2, N° 10) and, for each $x \in E$, $f_n(x) \to f_0(x)$. This implies that f_n $(n \geqslant 1)$ is simply bounded. The remainder follows from the proof of Corollary 5.

Corollary 5 will read in the particular case (when E, F are Banach spaces) as follows:

COROLLARY 6. *Let E and F be two Banach spaces, and T_n $(n \geqslant 1)$ a sequence of linear and continuous mappings of E into F. Then:*

(*a*) *$T_n(x)$ $(n \geqslant 1)$ is bounded for all $x \in E$ if and only if T_n $(n \geqslant 1)$ is uniformly bounded on $\|x\| \leqslant 1$.*

(*b*) *$T_n(x)$ converges to $T_0(x)$ for each $x \in E$ if and only if T_n is simply bounded and $T_n(x) \to T_0(x)$ on a dense subset E_0 of E.*

9. Duality theory

Definition 13. (*a*) Let E be an l.c. space. The set $E' = L(E, R)$ of all linear and continuous mappings of E into the reals R is called the *topological dual*, or simply, the *dual* of E. Clearly $E' \subset E^* \subset R^E$.

(*b*) The *value* of a linear functional $x' \in E'$ at $x \in E$ is denoted by $\langle x, x' \rangle$.

(1) (*a*) Clearly the mapping: $(x, x') \to \langle x, x' \rangle$ of $E \times E'$ into R is *bilinear* (i.e. linear in each variable). (*b*) A linear functional $x'\colon E \to R$ is continuous if and only if it is bounded on an open neighbourhood of 0.

(*c*) The *coarsest* locally convex topology for which the mapping: $x \to \langle x, x' \rangle$, for each $x' \in E'$, is continuous is called the *weak topology* $\sigma(E, E')$ on E. In the same way one defines the *weak* topology* $\sigma(E', E)$ as the coarsest one for which the mapping: $x' \to \langle x, x' \rangle$, for each $x \in E$, is continuous.

(2) $\sigma(E', E)$ is precisely the topology of simple convergence (§ 7) on E', which, in turn, is induced from the product topology defined on R^E.

(*d*) For each subset A of E, the set of all $x' \in E'$ such that $\langle x, x' \rangle \leqslant 1$, is called the *polar* of A and denoted by A^0. A^{00} is called the *bipolar*, i.e. the set of all $x \in E$ such that $\langle x, x' \rangle \leqslant 1$ for $x' \in A^0$.

The following properties of a polar are easy to verify:

PROPOSITION 12. *Let A, B, and A_α (α runs over an index set) be subsets of an l.c. space E.*

(*a*) *If $A \subset B$, then $A^0 \supset B^0$.*

(*b*) *If $\lambda \neq 0$, $(\lambda A)^0 = \lambda^{-1} A^0$.*

(*c*) $(\bigcup_\alpha A_\alpha)^0 = \bigcap_\alpha A_\alpha^0$.

(*d*) *For each set A, A^0 is a $\sigma(E', E)$-closed convex set containing* 0.

(*e*) *If A is circled, so is A^0.*

(*f*) *A^{00} = the convex closure (under $\sigma(E, E')$) of A and $\{0\}$.*

(*g*) $A^{000} = A^0$.

(*h*) *If A is a subspace of E, A^0 is a $\sigma(E', E)$-closed subspace of E'.* (*A^0 is called* orthogonal *to A*).

(i) *Let A be a subspace of E. $A^{00} = A$ if and only if A is $\sigma(E, E')$-closed. (The same is true for a subspace A of E'.)*

(3) Let M be a subspace of E. Then M^0 is a $\sigma(E', E)$-closed subspace of E'. Consider the quotient space E'/M^0 with the quotient topology. For each $x^0 \in E'/M^0$, $x^0 = x' + M^0$ $(x' \in E')$, and therefore, for each $x \in M$, $\langle x, x_1^0 \rangle = \langle x, x_2^0 \rangle$ if $x_1' - x_2' \in M^0$. Clearly $(x, x^0) \to \langle x, x^0 \rangle$ is a bilinear mapping of $M \times (E'/M^0)$. Therefore M and E'/M^0 are duals of each other.

Proposition 13. *Let M be a subspace of E and E' the dual of E. Then:*

(a) The weak topology $\sigma(M, E'/M^0)$ on M is the topology induced by $\sigma(E, E')$ on M.

(b) For the topology $\sigma(E'/M^0, M)$ to be equal to the quotient weak topology on E'/M^0, it is necessary and sufficient that M be $\sigma(E, E')$-closed in E.*

For the proof see [4] (Chapitre IV, § 1, N° 5, Propositions 6 and 7 respectively).

(4) Let E_α $(\alpha \in A)$ be an l.c. space and E'_α its dual. Then the dual of $\prod_\alpha E_\alpha$ is $\sum_\alpha E'_\alpha$, and the dual of $\sum_\alpha E_\alpha$ is $\prod_\alpha E'_\alpha$ ([4], Chapitre IV, § 1, Prop. 8).

Proposition 14. *A set M' in E', the dual of an l.c. space E_u, is equicontinuous if and only if there exists a u-neighbourhood U of 0 in E_u such that $M' \subset U^0$.*

Proof. If M' is equicontinuous, then for $V = \{\lambda\colon |\lambda| < 1\}$, $\bigcap_{x' \in M'} x'^{-1}(V)$ is a u-neighbourhood U of 0 in E_u (Def. 12). In other words, for all $x' \in M'$, $\langle x, x' \rangle \leqslant 1$, where $x \in U$. This implies that $M' \subset U^0$. On the other hand, for any subset M' in E', $M' \subset U^0$ implies $M'^0 \supset U^{00} \supset U$. This clearly shows that M' is equicontinuous.

Definition 14. Let u and v be two locally convex topologies on a vector space E. u and v are said to be *equivalent* (or in notation $u \sim v$) if E_u and E_v have the same topological dual, i.e. $E'_u = E'_v$.

In the Bourbaki sense u is said to be *compatible with the duality* between E_v and E'_v if $E'_u = E'_v$.

For an l.c. space E_u, $\sigma(E, E') \sim u$.

Proposition 15. *Let u and v be two equivalent locally convex topologies on a vector space E. Then a convex subset M of E is u-closed if and only if it is v-closed.*

Proof. Since u and v are equivalent, $E'_u = E'_v$. Further, since the weak topology of an l.c. space E_u is determined by its dual, therefore the weak topologies with respect to u and v are the same. Now observe that in order to prove the proposition, it is sufficient to establish that a convex set is u-closed if and only if it is σ-closed, because by symmetry it will follow that v-closed convex sets are the same as σ-closed convex sets.

Observe that every convex u-closed set C is the intersection of all closed half-spaces containing C. But every half-space is $\{x: x \in E, \langle x, x' \rangle \leqslant \alpha, \alpha \text{ real}, x' \in E'_u\}$. Since every linear functional which is σ-continuous is u-continuous and conversely, the proposition follows immediately.

For a deeper result showing that even bounded sets are the same under equivalent topologies, we consider the following:

THEOREM 6. (*Mackey.*) *Let E be an l.c. space and E' its dual. A locally convex topology u on E is equivalent to $\sigma(E, E')$ if and only if u is an $\mathfrak{S}$-topology, where $\mathfrak{S}$ consists of $\sigma(E', E)$-compact, convex, circled sets M_α of E' such that $\bigcup_\alpha M_\alpha = E'$.*

Proof. Suppose $u \sim \sigma(E, E')$, i.e. $E'_u = E'_\sigma$. Consider $\mathfrak{S} = \{U^0_\alpha\}$, where U_α runs over a fundamental system of convex, closed, circled u-neighbourhoods of 0 in E. Each U^0_α is convex and circled; and it is also $\sigma(E', E)$-compact, as follows by Tychonoff's theorem. Clearly $\bigcup_\alpha U^0_\alpha = E'$, because $\bigcap_\alpha U_\alpha = \{0\}$. Further, $U^{00}_\alpha = U_\alpha$ shows that u is an $\mathfrak{S}$-topology as described in the theorem.

Conversely, let $\mathfrak{S} = \{M_\alpha\}$, where M_α is a convex, circled, and $\sigma(E', E)$-compact set in E' such that $\bigcup_\alpha M_\alpha = E'$. By definition (§ 7), the $\mathfrak{S}$-topology u defined by the given family $\mathfrak{S} = \{M_\alpha\}$ is a Hausdorff locally convex topology on E. We want to show that $u \sim \sigma$. Observe that u remains unchanged if one enlarges $\mathfrak{S}$ by including the convex hulls of finite unions of M_α's, because the convex hull of a finite number of compact sets is compact (Prop. 1, § 4). Since a finite subset of E' is $\sigma(E', E)$-compact, therefore $\mathfrak{S}$ also includes the convex, circled hulls of finite subsets of E'. In other words, $u \supset \sigma(E, E')$. From this it follows that $E'_u \supset E'_\sigma$. Now in order to complete the proof of the theorem, we show that $E'_\sigma \supset E'_u$. Let $x' \in E'_u$. Since x' is continuous on E_u, there exists an M_α in $\mathfrak{S}$ such that $\langle M^0_\alpha, x' \rangle \leqslant 1$. But this shows that x' belongs to the polar N_α of M^0_α in E'_u. Since $\sigma(E'_u, E_u)$ induces $\sigma(E'_\sigma, E_\sigma)$ on E'_σ, and the polar of any subset of E in E'_u is the closure (under $\sigma(E'_u, E_u)$) of the polar of the same set in E'_σ (Prop. 12 (f)), M_α is $\sigma(E'_u, E_u)$-compact and therefore $\sigma(E'_u, E_u)$-closed. Hence $N_\alpha = M_\alpha$. This proves that $x' \in E'_\sigma$ and therefore $E'_\sigma \supset E'_u$.

THEOREM 7. *Let E_u be an l.c. space and E' its dual. Then the following statements are equivalent:*

(*a*) *E_u is barrelled.*

(*b*) *Each $\sigma(E', E)$-bounded subset of E' is equicontinuous.*

Proof. (*a*) ⇒ (*b*) follows as a particular case from Theorem 5, § 8, when F is the set of real numbers.

For (*b*) ⇒ (*a*), let B be a barrel in E_u. Since B is absorbing, B^0 is $\sigma(E', E)$-bounded. Therefore (*b*) implies B^0 is equicontinuous. Hence by Proposition 14, $B^{00} \supset U$. Since B is convex, circled, and u-closed (and therefore σ-closed by Prop. 15), $B = B^{00} \supset U$ proves that B is a neighbourhood of 0. In other words, E_u is a barrelled space.

COROLLARY 7. *Let E_u be a t-space. Then E', with $\sigma(E', E)$, is quasi-complete.*

Proof. It is easy to see that every equicontinuous set in E' is relatively compact. If M' is a $\sigma(E', E)$-bounded set of E', then M', being equicontinuous by Theorem 7, is relatively compact. Now if in addition M' is closed, then it, being compact, is complete in the uniform structure induced on M' from E'. This proves the corollary.

COROLLARY 8. *If E_u is a t-space, then, in E', the following sets are the same*: (*a*) *equicontinuous*, (*b*) *relatively compact*, (*c*) *$\sigma(E', E)$-bounded.*

Proof. (*a*) and (*c*) are the same due to Theorem 7. (*c*) ⇒ (*b*) due to Corollary 7, and the converse, i.e. (*b*) ⇒ (*c*) is trivially true.

Definition 15. (*a*) The $\mathfrak{S}$-topology on an l.c. space E, where $\mathfrak{S}$ consists of all convex, circled, $\sigma(E', E)$-compact subsets of the dual E' of E, is called the *Mackey topology* and denoted by $\tau(E, E')$.

(*b*) An l.c. space E_u is said to be a *Mackey space* if $u = \tau(E, E')$.

(5) (*a*) An l.c. space E_u is a Mackey space if and only if each convex relatively compact subset of E', in the $\sigma(E', E)$-topology, is equicontinuous. (*b*) A t-space is a Mackey space.

THEOREM 8. (*Mackey.*) *Let E_u be an l.c. space and E' its dual. Then a subset M of E is $\sigma(E, E')$-bounded if and only if it is $\tau(E, E')$-bounded.*

Proof. Clearly $\sigma(E, E') \subset u \subset \tau(E, E')$. Therefore by continuity of the identity mapping: $E_\tau \to E_\sigma$, if M is τ-bounded then it is σ-bounded. The converse follows from Theorem 4, § 7, because $\tau(E, E')$ is an $\mathfrak{S}$-topology where $\mathfrak{S}$ consists of all $\sigma(E', E)$-compact convex circled sets of E'.

Definition 16. (*a*) Let E' be the dual of an l.c. space E_u. The following notations will be used for $\mathfrak{S}$-topologies other than $\sigma(E', E)$ on E':

c = the $\mathfrak{S}$-topology on E' where $\mathfrak{S}$ consists of all convex u-compact sets of E.

k = the $\mathfrak{S}$-topology on E' where $\mathfrak{S}$ consists of all u-compact sets of E.

p = the $\mathfrak{S}$-topology on E' where $\mathfrak{S}$ consists of all u-precompact sets of E.

β = the $\mathfrak{S}$-topology on E' where $\mathfrak{S}$ consists of all u-bounded sets of E. It is also called the *strong topology*.

Observe that β remains unchanged if $\mathfrak{S}$ consists of all σ-bounded sets instead of u-bounded sets, due to Theorem 8.

(6) Each of these topologies is locally convex, and moreover,

$$\sigma(E', E) \subset c \subset k \subset p \subset \beta.$$

(*b*) The set E' with the β-topology is called the *strong dual* and denoted by E'^{β}. The set E' without mention of any topology will be understood as the *weak dual* E', i.e. E' endowed with $\sigma(E', E)$. The set E' with an l.c. topology v other than $\sigma(E', E)$ will be denoted by E'^{v}.

(*c*) The dual $E'^{\beta'}$ of E'^{β} is called the *bidual* of E.

(*d*) If $E'^{\beta'} = E$ (algebraically) then E is said to be *semi-reflexive.*

(*e*) If $E'^{\beta'\beta} = E_u$ (topologically) then E is said to be *reflexive.*

THEOREM 9. *Let E_u be an l.c. space and E' its dual. The following statements are equivalent:*

(*a*) *E_u is quasi-barrelled.*

(*b*) *Every β-bounded set of E' is equicontinuous.*

Proof. (*a*) $\Rightarrow$ (*b*): Let M be a β-bounded set of E'. Then M is σ-bounded because $\beta \supset \sigma(E', E)$. Hence M^0 is a barrel of E_u, and M being β-bounded, M^0 is bornivorous. Therefore M^0 is a u-neighbourhood of 0 because of (*a*). In other words, M is equicontinuous due to Proposition 14.

(*b*) $\Rightarrow$ (*a*): Let B be a barrel which is bornivorous. Then B^0 is β-bounded, and hence, by (*b*), equicontinuous. Therefore $B^{00} = B$ (because B is a barrel) is a u-neighbourhood of 0. This completes the proof.

PROPOSITION 16. *A quasi-barrelled space E_u is a Mackey space.*

Proof. Clearly u-bounded and $\tau(E, E')$-bounded sets are the same (Theorem 8). Therefore $E_u'^{\beta} = E_\tau'^{\beta}$. Let M be a convex compact set of $E_\tau'^{\beta}$. Then M is $\sigma(E', E)$-compact because $\beta \supset \sigma(E', E)$ and therefore equicontinuous when E is endowed with $\tau(E, E')$. Now M, being equicontinuous, is β-bounded and therefore equicontinuous when E is endowed with the initial topology u, due to Theorem 9. Thus, by ((5) (a)), E_u is a Mackey space.

COROLLARY 9. *Every bornological space (in particular a metrizable l.c. space) is a Mackey space.*

Proof. This follows from the fact that every bornological space is quasi-barrelled.

(7) (a) If E is bornological then E'^{β} is complete. (b) In particular, if E is a metrizable l.c. space then E'^{β} is complete.

(8) (a) An l.c. space E_u is semi-reflexive if it is reflexive. (b) If E_u is reflexive then $E_u'^{\beta}$ is also reflexive. (c) An l.c. space E_u is semi-reflexive if and only if each $\sigma(E, E')$-closed and σ-bounded set is $\sigma(E, E')$-compact. (d) An l.c. space E_u is reflexive if and only if E_u is a t-space and each $\sigma(E, E')$-bounded set is relatively $\sigma(E, E')$-compact.

Definition 17. An l.c. space E_u which is a t-space and in which each bounded set is relatively compact, is called a *Montel* space.

(9) Every Montel space is reflexive.

PROPOSITION 17. *If E_u is a Montel space then $E_u'^{\beta}$ is also a Montel space.*

Proof. Since E'^{β} is reflexive due to ((8) (b)), therefore by ((8) (d)), E'^{β} is a t-space. Now we show that each β-bounded and β-closed set B of E'^{β} is β-compact. Since E_u is a t-space (because E_u is reflexive), B, being σ-bounded (because $\beta \supset \sigma$), is equicontinuous. Hence on B, the topologies k and $\sigma(E', E)$ coincide (Prop. 11, § 8). But $k = \beta$ because each u-closed and u-bounded set of E_u is u-compact by hypothesis. Therefore β coincides with $\sigma(E', E)$ on B. Now in view of the fact that B is given to be β-closed, B is σ-compact and therefore β-compact.

Let E_u and F_v be two l.c. spaces. Let f be a mapping of E_u into F_v. Let E_u', F_v' denote the duals of E_u and F_v respectively. Then there exists a *transpose* mapping f' of F_v' into E_u', defined as

$$\langle f(x), y' \rangle = \langle x, f'(y') \rangle,$$

$x \in E$, $y' \in F'$.

(10) (a) If f is continuous so is f'. For $f(E)$ to be weakly dense in F, it is necessary and sufficient that f' be 1:1, and for f to be weakly open it is necessary and sufficient that $f'(F')$ be $\sigma(E', E)$-closed in E', provided f is already given to be continuous in both cases.

(b) If $f: E_u \to F_v$ is continuous then $f: E_\sigma \to F_{\sigma'}$ is also continuous where $\sigma = \sigma(E, E')$, $\sigma' = \sigma(F, F')$, but not conversely. The converse is true if E is a Mackey space or in particular a t-space.

Proposition 18. *Let E_u and F_v be two l.c. spaces and let f be a linear almost continuous mapping of E_u into F_v. If f is a continuous mapping of E_σ into F_v ($\sigma = \sigma(E, E')$) then it is also a continuous mapping of E_u into E_v.*

Proof. Let V be a convex closed neighbourhood of 0 in F_v. Then $f^{-1}(V)$ is a closed convex σ-neighbourhood of 0 in E. But

$$f^{-1}(V) = uf^{-1}(V)$$

(by Proposition 15) which is a u-neighbourhood of 0 in E_u because f is almost continuous.

Proposition 19. *Let E_u and F_v be two l.c. spaces and f a linear continuous almost open mapping of E_u onto F_v. If the image of F', under the transpose mapping f': $F' \to E'$, is $\sigma(E', E)$-closed, then f is open.*

Proof. In view of the above remarks ((10) (a)), $f'(F')$ being $\sigma(E', E)$-closed in E' implies f: $E_\sigma \to F_{\sigma'}$ is open, where $\sigma = \sigma(E, E')$ and $\sigma' = \sigma(F, F')$. We wish to apply Ex. 3 (a) ([4], Chapitre IV, § 4, N° 2). Let M' be an equicontinuous subset of E' which is contained in $f'(F')$. By Proposition 14 there exists a u-neighbourhood U of 0 in E such that $M' \subset U^0$. Whence we have

$$f'^{-1}(M') \subset f'^{-1}(U^0 \cap f'(F')) = f'^{-1}(U^0) = (f(U))^0 = (\overline{f(U)})^0.$$

But since f is almost open, $\overline{f(U)}$ is a v-neighbourhood of 0 in F and hence $(\overline{f(U)})^0$ is equicontinuous in F'. Therefore f is open.

If F_v, in the above proposition, is a t-space then one can drop almost openness of f by Prop. 6, § 6.

3

THE OPEN-MAPPING AND CLOSED-GRAPH THEOREMS

1. Finite-dimensional TVS's and the open-mapping theorem

LET E and F be two TVS's and consider the following statements:

(A) *A linear and continuous mapping f of E onto F is open.*

(B) *A linear mapping g of F into E with the closed graph is continuous.*

The statements (A) and (B) are not true for every pair of TVS's E and F, as the reader can very easily convince himself by counter-examples. If they are true for some TVS's E and F with or without any additional conditions on the mappings, (A) and (B) are called the open mapping and closed graph theorems respectively.

PROPOSITION 1. *Let f be a linear mapping of a TVS E into another TVS F. If f is continuous then the graph of f is closed.*

Proof. Let $G = \{(x, f(x)), x \in E\}$ be the graph of f in $E \times F$, and let $(x, y) \in \bar{G}$. Hence, for each closed neighbourhood V of 0 in F,

$$f^{-1}(y+V) \cap (x+U) \neq \emptyset$$

for any arbitrary neighbourhood U of 0 in E. That means there exists $x_u \in x+U$ such that $f(x_u) \in y+V$. Since U is arbitrary, $x_u \to x$; and hence, by the continuity of f, it follows that $f(x) \in y+V$ because V is closed. Now since V is arbitrary, $f(x) = y$ shows that $(x, y) \in G$. This proves the proposition.

The converse of Proposition 1 is not true as is easily seen.

COROLLARY 1. *Let f be a linear,* 1:1, *and open mapping of a TVS E onto another TVS F. Then the graph G of f is the same as that of f^{-1}, and G is also closed.*

Proof. Since f is 1:1 and onto, f^{-1} exists. Moreover, $f(x) = y$ if and only if $f^{-1}(y) = x$ and hence the graphs are the same. Further, openness of f is equivalent to continuity of f^{-1} under the given conditions. Therefore by Proposition 1, the corollary follows.

(1) (*a*) Since the composition of two continuous mappings is continuous, the graph of the composition of two continuous mappings is closed. (*b*) In general, however, the graph of the composition of two

mappings, each having the closed graph, is *not* closed, as is easy to see. (*c*) Further, the set of all mappings of a TVS E into another TVS F, each having the closed graph, does *not* necessarily form a vector space, e.g. see [37], p. 45, 2.12.

PROPOSITION 2. *Let E be a real TVS of dimension n. Then E is homeomorphic to R^n* (Chapter 2, § 1, Ex. 2).

Proof. We prove the proposition by induction. Let $n = 1$. For each $\lambda \in R$ and a fixed point x_0 $(\neq 0) \in E$, consider the mapping $\varphi: \lambda \to \lambda x_0$. Clearly φ is a linear, continuous, and one-to-one mapping of R onto E. Let $U_\epsilon = \{\lambda: |\lambda| < \epsilon, \epsilon > 0\}$ be a neighbourhood of 0 in R, and $0 < \lambda_0 \in U_\epsilon$. E being Hausdorff, there exists a circled neighbourhood V of 0 such that $\lambda_0 x_0 \notin V$. We wish to show that $\varphi(U_{\lambda_0}) \supset V$. If not, then for some λ, $\lambda x_0 \in V$ implies $|\lambda| \geqslant \lambda_0$ or $|\lambda^{-1}|\lambda_0 \leqslant 1$. Since V is circled, it follows that $\lambda_0 x_0 = \lambda_0 \lambda\lambda^{-1} x_0 \in V$, which gives a contradiction. Therefore φ is open.

For the general case, let $e_1, \dots, e_n$ be a basis of E and consider the mapping $\varphi: (\lambda_1, \dots, \lambda_n) \to \sum_{i \leqslant n} \lambda_i e_i$ of R^n into E. By induction on n, it can be assumed that the proposition is true for $n-1$, i.e. R^{n-1} is homeomorphic to a vector subspace H (of E) generated by $e_1, \dots, e_{n-1}$. Since R^{n-1} is complete, H is a closed subspace of E and $E/H = Re_n$, as is easily seen. Since R is homeomorphic to Re_n, R^n is homeomorphic to $H \times (E/H)$. But the latter is equal to E. This completes the proof.

COROLLARY 2. *Let E and F be any two TVS's of dimension n. Then E and F are homeomorphic.*

Proof. Since each of E and F is homeomorphic to R^n by Proposition 2, therefore E and F are homeomorphic to each other.

COROLLARY 3. *Every finite-dimensional vector subspace H of a TVS E is closed.*

Proof. Since H is homeomorphic to R^n, where n is the dimension of H, H is complete (because R^n is complete), and hence H is closed in E.

(2) Let E, F, and G be three TVS's. Let f be a linear mapping of E onto F and g a linear mapping of F into G. Let $h = g \circ f$ be the linear mapping of E into G. Suppose f is continuous and open. Then: (*a*) g is continuous if and only if h is continuous. (*b*) g is open if and only if h is open. (*c*) g is onto if and only if h is onto.

THEOREM 1. *Let F be a finite-dimensional TVS and E any TVS. Then:*

(*a*) *A linear and continuous mapping f of E onto F is open.*

(*b*) *A linear mapping g of F into E is continuous.*

Proof. (*a*) Clearly $f^{-1}(0)$ is a closed subspace of E. Let $\varphi: E \to E/f^{-1}(0)$ and $f = f_1 \circ \varphi$, where $f_1: E/f^{-1}(0) \to F$. Since φ is continuous and open, in order to establish (*a*) it is sufficient to show that f_1 is open. Observe that $E/f^{-1}(0)$ and F are TVS's having the same dimension and therefore are homeomorphic to each other by Corollary 2. Hence f is open.

(*b*) Observe that each linear mapping of R^n (where n is the dimension of F) into E is of the form: $(\lambda_1,\ldots,\lambda_n) \to \sum_{i \leqslant n} \lambda_i x_i$ $(x_i \in E)$ and therefore continuous. Now the fact that F is homeomorphic to R^n establishes (*b*).

2. Banach's theorems

The open-mapping and closed-graph theorems of Banach for F-spaces will be derived from more general theorems of this section. First of all we have the following:

LEMMA 1. *Let U_n $(n \geqslant 1)$ denote a countable fundamental system of neighbourhoods of* 0 *in a metrizable TVS E and let F be a TVS. If f is a linear, continuous, and almost open mapping of E into F then*

$$\bigcap_{n=1}^{\infty} \overline{f(U_n)} = \{0\}.$$

Proof. Observe that U_n $(n \geqslant 1)$ can be assumed to be decreasing, i.e. $U_{n+1} \subset U_n$ for each n. By almost openness of f for each U_n, $\overline{f(U_n)}$ is a neighbourhood of 0 in F. Let $y \in \bigcap_{n=1}^{\infty} \overline{f(U_n)}$ and let V be a circled and closed neighbourhood of 0 in F. Then $(y+V) \cap f(U_n) \neq \emptyset$ for each n. Let $x_n \in U_n$ such that $f(x_n) \in y+V$ or $f(x_n)-y \in V$ for all $n \geqslant 1$. Since U_n $(n \geqslant 1)$ forms a decreasing sequence of a fundamental system of neighbourhoods of 0, $x_n \in U_n$ implies $x_n \to 0$ as $n \to \infty$. Since f is continuous and linear, $f(x_n) \to 0$ and thus $-y \in V$ (because V is closed). Since V is circled and arbitrary, therefore $y = 0$.

THEOREM 2. (*Open mapping.*) *Let E be an F-space and F a TVS. Then a linear, continuous, and almost open mapping of E into F is open.*

Proof. Let U_n $(n \geqslant 1)$ be a decreasing sequence of a fundamental system of closed neighbourhoods of 0 in E such that $U_{n+1}+U_{n+1} \subset U_n$ for each n. In order to prove the theorem it is sufficient to show that $f(U_n)$ is a neighbourhood of 0 in F for each n. Let $\overline{f(U_n)} = W_n$. Then W_n is a neighbourhood of 0 in F for each n because f is almost open. We shall show that $f(U_k) \supset W_{k+1}$ for each k.

Let $y \in W_{k+1}$. Since $f(U_{k+1})$ is dense in W_{k+1}, there exists y_1 with preimage $x_1 = f^{-1}(y_1) \in U_{k+1}$ such that $y-y_1 \in W_{k+2}$. Inductively proceeding, we assume that there exists y_n with pre-image $x_n = f^{-1}(y_n) \in U_{k+n}$ such that

$$y - \sum_{j=1}^{n} y_j \in W_{k+n+1}. \tag{2.1}$$

Since $$\sum_{i=0}^{p} x_{n+i} \in \sum_{i=0}^{p} U_{k+n+i}, \quad \text{and} \quad U_{k+n} \supset \sum_{i=1}^{p} U_{k+n+i},$$

therefore $$\sum_{i=0}^{p} x_{n+i} \in U_{k+n} + U_{k+n} \subset U_{k+n-1}. \tag{2.2}$$

Since U_n $(n \geqslant 1)$ forms a decreasing sequence of a fundamental system of neighbourhoods of 0 in E, for each arbitrary neighbourhood V of 0 in E, there exists a positive integer n_0 such that for all $n \geqslant n_0$, $U_{k+n-1} \subset V$. Therefore $\sum_{i=0}^{p} x_{n+i} \subset V$ for all $n \geqslant n_0$, $p \geqslant 0$. This shows that the partial sums of the series $\sum_{i=1}^{\infty} x_i$ form a Cauchy sequence. But E being complete implies there exists $x \in E$ such that

$$x = \sum_{i=1}^{\infty} x_i.$$

From (2.2), by putting $n = 1$, we have

$$\sum_{i=0}^{p} x_{i+1} \in U_k \quad \text{for each } p \geqslant 0.$$

Since U_k is closed so, letting $p \to \infty$, we have

$$x = \sum_{i=1}^{\infty} x_i \in U_k. \tag{2.3}$$

And the continuity of f implies

$$f(x) = \sum_{i=1}^{\infty} f(x_i) = \sum_{i=1}^{\infty} y_i. \tag{2.4}$$

From (2.1) follows that

$$y - \sum_{i=1}^{n+p} y_i \in W_{k+n+p+1} \subset W_{k+n+1} \quad (p \geqslant 0).$$

Since p is arbitrary, therefore

$$y - \sum_{i=1}^{m} y_i \in W_{k+n+1} \quad \text{for large } m.$$

And W_{k+n+1} being closed we have (letting $m \to \infty$)

$$y - \sum_{i=1}^{\infty} y_i \in W_{k+n+1} \quad \text{for each } n.$$

Therefore $$y - f(x) = y - \sum_{i=1}^{\infty} y_i \in \bigcap_{n=1}^{\infty} W_{k+n+1}.$$

By Lemma 1, we have $y-f(x) = 0$. But since from (2.3) $x \in U_k$, therefore

$$y = f(x) \in f(U_k).$$

This completes the proof.

If, in addition, one assumes E and F in Theorem 2 to be locally convex, the proof obviously remains valid. Hence, in view of Proposition 6, Chapter 2, § 6, we have the following:

THEOREM 3. *Let E be a Fréchet space and F a barrelled l.c. space. Then a linear continuous mapping of E onto F is open.*

An immediate consequence of the above theorem is:

COROLLARY 4. *Let E_u be a Fréchet space and v any l.c. topology coarser than u on E such that E_v is barrelled. Then $u = v$.*

The following lemmas are needed to prove Theorem 4:

LEMMA 2. *Let F be a TVS and E a metrizable TVS with a countable fundamental system of neighbourhoods V_n ($n \geqslant 1$) of 0. Let f be a 1:1 linear almost continuous mapping of F into E, the graph of which is closed in $F \times E$. Then*

$$\bigcap_{n=1}^{\infty} \overline{f^{-1}(V_n)} = \{0\}.$$

Proof. We can assume that, for each n, $V_{n+1}+V_{n+1} \subset V_n$. Therefore V_n ($n \geqslant 1$) is a decreasing sequence of a fundamental system of neighbourhoods of 0 in E. Since f is almost continuous, $U_n = \overline{f^{-1}(V_n)}$ is a neighbourhood of 0 in F for each n.

Let $x \in U_n$ for each n, and let $U \times V$ be an arbitrary neighbourhood of 0 in $F \times E$. By the definition of U_n, $(x+U) \cap f^{-1}(V_n) \neq \emptyset$ for each n. Hence there exists $x_n = f^{-1}(y_n)$ ($y_n \in V_n$) such that $x_n \in x+U$ for each n. Since $\{V_n\}$ is a decreasing sequence of a fundamental system of neighbourhoods of 0 in E, there exists a large positive integer m such that $V_m \subset V$. Hence there exists $x_m = f^{-1}(y_m)$ ($y_m \in V_m$) such that

$$x_m \in x+U \quad \text{and} \quad f(x_m) = y_m \in V.$$

Therefore

$$(x_m, f(x_m)) \in G \cap ((x+U) \times V).$$

This shows that $(x, 0) \in \overline{G} = G$. In other words, $f(x) = 0$. Since f is one-to-one, $x = 0$.

LEMMA 3. *Let F and E be two TVS's and g a linear mapping of F into E, the graph of which is closed in $F \times E$. Then $g^{-1}(0)$ is a closed subspace of F.*

Proof. Let $H = \{x : g(x) = 0\}$. Clearly H is a subspace. Let $x \in \overline{H}$ and consider $(x, 0)$. Let U and V be arbitrary neighbourhoods of 0 in F and E respectively. There exists, indeed, a point $x_1 \in x+U$ such that

$g(x_1) = 0 \in V$. Therefore $(x, 0) \in \bar{G} = G$. In other words, $g(x) = 0$ and hence $x \in H$.

PROPOSITION 3. *Let E and F be two TVS's and f a linear mapping of E into F so that $f^{-1}(0)$ is a closed subspace of E. Let f_1: $E/f^{-1}(0) \to F$ and φ: $E \to E/f^{-1}(0)$. Then*

(a) f_1 is almost open if and only if f is almost open.

(b) f_1 is almost continuous if f is so.

Proof. Observe that $f = f_1 \circ \varphi$. Hence for any neighbourhood U of 0 in E,

$$\overline{f(U)} = \overline{f_1(\varphi(U))}.$$

Since φ is open, (a) follows immediately.

Further, if V is a neighbourhood of 0 in F,

$$\overline{f_1^{-1}(V)} = \overline{\varphi(f^{-1}(V))} \supset \varphi(\overline{f^{-1}(V)}),$$

because φ is continuous. Now if f is almost continuous so is f_1, and thus (b) is established.

(1) Let g be a linear mapping of a TVS F into another TVS E. Let φ be the canonical mapping: $F \to F/g^{-1}(0)$. Let g_1: $F/g^{-1}(0) \to E$ such that $g = g_1 \circ \varphi$. Then the graph of g is closed if and only if the graph of g_1 is closed.

THEOREM 4. (*Closed graph.*) *Let F be a TVS and E an F-space. Let g be a linear mapping of F into E, the graph G of which is closed in $F \times E$. If g is almost continuous then g is continuous.*

Proof. In view of Lemma 3, Proposition 3 and (1), it can be assumed, in addition, that g is 1:1. Let $\{V_n\}$ be a countable fundamental system of closed circled neighbourhoods of 0 in E such that $V_{n+1} + V_{n+1} \subset V_n$ for each n. Since g is almost continuous, for each V_k, $U_k = \overline{g^{-1}(V_k)}$ is a neighbourhood of 0 in F. In order to prove the theorem we shall show that $g(U_{k+1}) \subset V_k$ for each k.

Let $y \in g(U_{k+1})$, i.e. $y = g(x)$ $(x \in U_{k+1})$. Since $g^{-1}(V_{k+1})$ is dense in U_{k+1}, there exists $x_1 = g^{-1}(y_1)$, $y_1 \in V_{k+1}$ such that $x - x_1 \in U_{k+2}$. Inductively proceeding, there exists $x_n = g^{-1}(y_n)$ $(y_n \in V_{k+n})$ such that $x - \sum_{i \leqslant n} x_i \in U_{k+n+1}$. Clearly

$$\sum_{n \leqslant i \leqslant m} y_i \in \sum_{n \leqslant i \leqslant m} V_{k+i} \subset V_{k+n-1} \quad (m \geqslant n).$$

Since V_n $(n \geqslant 1)$ is a decreasing sequence of a fundamental system of neighbourhoods of 0, for an arbitrary neighbourhood V of 0 there exists $n_0 > 0$ such that $V_{k+n_0-1} \subset V$. Hence the partial sums of the series $\sum_{i=1}^{\infty} y_i$, being a Cauchy sequence implies there exists y' such that

$$y' = \sum_{i=1}^{\infty} y_i = \sum_{i=1}^{\infty} g(x_i),$$

because E is an F-space. Since $\sum_{i \leqslant m} y_i \in V_k$ for each m, and V_k is closed, therefore $y' \in V_k$.

Further, $x - \sum_{i=1}^{n} x_i \in U_{k+n+1}$ implies

$$x - \sum_{i=1}^{n+p} x_i \in U_{k+n+p+1} \subset U_{k+n+1} \quad (p \geqslant 0).$$

p being arbitrary, $x - \sum_{i=1}^{\infty} x_i \in U_{k+n+1}$ for each n because U_{k+n+1} is closed. Therefore

$$x - \sum_{i=1}^{\infty} x_i \in \bigcap_{n=1}^{\infty} U_{k+n+1}.$$

Hence, by Lemma 2, $x - \sum_{i=1}^{\infty} x_i = 0$. Clearly

$$\Big(x - \sum_{i \leqslant n} x_i,\ g(x) - \sum_{i \leqslant n} g(x_i)\Big) \in G.$$

Since G is closed, $(0, g(x) - y') \in G$. Therefore $y = g(x) = y' \in V_k$. This proves the theorem.

THEOREM 5. *Let E be a Fréchet space and F a barrelled l.c. space. Let g be a linear mapping of F into E, the graph of which is closed in $F \times E$. Then g is continuous.*

Proof. Since F is barrelled, by Proposition 6 (Chapter 2, § 6) g is almost continuous. Hence by Theorem 4, g is continuous.

For 'onto' mappings we can prove Theorem 2 under weaker conditions. Actually then we can derive Theorem 2 from Theorem 4, as we see in the sequel.

THEOREM 6. *Let E be an F-space and F any TVS. Let f be a linear mapping of E* onto *F so that the graph of f is closed in $E \times F$. If f is almost open then f is open.*

Proof. In view of Lemma 3, Proposition 3, (1), and the fact that a quotient space of an F-space is an F-space, it can be assumed, in addition, that f is 1:1. Hence $f^{-1} = g$ exists. Since the graphs of f and g are the same, the graph of g is closed; and g is almost continuous because f is almost open. Then by Theorem 4, g is continuous or in other words, f is open.

Remark. Now the fact that the continuous mappings have closed graphs implies Theorem 6 is more general than Theorem 2 for 'onto' mappings, and Theorems 4 and 5 imply Theorems 2 and 3, respectively.

Corollary 5. *Let E be a Fréchet space and F a barrelled space. Let f be a linear mapping of E onto F so that the graph of f is closed. Then f is open.*

Proof. This is immediate from Theorem 6.

It is easy to see that if E is any TVS and F a Baire TVS then a linear mapping f of E onto F is almost open ([3], Chapitre I, § 3, N° 3, Lemma 1), and a linear mapping of F into E is almost continuous. Hence from Theorems 2 and 4 we have the following:

Theorem 7. (*Banach.*) *Let E be an F-space and F a Baire TVS. Then:*

(*a*) *Every linear and continuous mapping of E onto F is open.*

(*b*) *Every linear mapping g of F into E, the graph of which is closed in $F \times E$, is continuous.*

In view of Baire's theorem (Chapter 1, § 6), the following is a particular case of Theorem 7.

Theorem 8. *Let E and F be any two F-spaces. Then:*

(*a*) *A linear and continuous mapping of E onto F is open.*

(*b*) *A linear mapping of F into E with the closed graph is continuous.*

Lemma 4. *Let E and F be two TVS's. Let f be a linear mapping of E into F such that $f(E)$ is of the second category in F. Then f is almost open.*

Proof. For each neighbourhood U of 0 in E, $\bigcup_{n \geqslant 1} \alpha^n U = E$ $(\alpha > 1)$ implies $f(E) = \bigcup_{n \geqslant 1} \alpha^n f(U)$. Hence for some n, $\overline{\alpha^n f(U)}$ is a neighbourhood of 0 in F. From this it follows that $\overline{f(U)}$ itself is a neighbourhood of 0 in F.

Corollary 6. *Let E be an F-space and F a TVS. Let f be a linear continuous mapping of E into F such that $f(E)$ is of the second category in F. Then f is open and onto.*

Proof. The fact that $f(E)$ is of the second category in F implies f is almost open, by Lemma 4. Therefore f is open by Theorem 2. Since each neighbourhood of 0 is absorbing, it follows immediately that f is onto.

Corollary 7. (*Banach* [1].) *Let E be an F-space and f a linear continuous mapping of E into any TVS F. Then either $f(E)$ is of the first category in F or $f(E) = F$.*

Proof. This follows from Corollary 6.

COROLLARY 8. *Let E and F be two F-spaces. Then every linear* 1:1 *continuous mapping of E onto F is a homeomorphism.*

Proof. This is immediate from Theorem 8.

Another version of this corollary is the following:

COROLLARY 9. *Let E_u be any F-space and v any topology on E such that E_v is an F-space and either $u \supset v$ or $v \supset u$. Then $u = v$.*

3. Some generalizations of Theorems 3 and 5

In this section all TVS's will be locally convex. As pointed out earlier, clearly all results of § 2 remain true if one assumes all vector spaces involved to be locally convex.

Dieudonné and Schwartz [9] proved statement (A) (§ 1) for more general spaces, namely LF-spaces. Köthe [20] established statements (A) and (B) for generalized LF-spaces (Chapter 2, § 5, Def. 8 (d)). Grothendieck [12] proved statements (A) and (B) in a form of such generality that it includes theorems of Dieudonné and Schwartz and that of Köthe as well. We shall derive these results from a more general result (Theorem 9) in this section. The proof of Theorem 9 is the arrangement of the proof of a still more general theorem due to A. P. Robertson and W. Robertson [31] which is given in Chapter 4.

LEMMA 5. *Let E be an l.c. space and H a vector subspace of E such that H is of the second category in E. Then H is barrelled in the induced topology.*

Proof. Let B be a barrel in H. Then $H \subset \bigcup_{n \geqslant 1} n\bar{B}$ and therefore $\bar{B}$ is a neighbourhood of 0 in E. Hence $\bar{B} \cap H = B$ (because B is closed in the induced topology) is a neighbourhood of 0 in H and therefore H is barrelled.

THEOREM 9. *Let E be a generalized LF-space and F an inductive limit of Baire l.c. spaces. Then:*

(a) A linear continuous mapping f of E onto F is open.

(b) A linear mapping g of F into E with the closed graph is continuous.

Proof. First we prove part (b). Let E_n $(n \geqslant 1)$ be a defining sequence of Fréchet spaces for E with mappings f_n $(n \geqslant 1)$ (Chapter 2, § 5, Def. 8 (b)). Let F_α $(\alpha \in A)$ denote a family of Baire l.c. spaces for F with mappings g_α (Chapter 2, § 5, Def. 8 (a)). Let h_α denote the composition mapping $g \circ g_\alpha \colon F_\alpha \to E$. The graph G_α of h_α in $F_\alpha \times E$, being the inverse image of the graph G of g in $F \times E$ under the mapping $F_\alpha \times E \to F \times E$ which is continuous (by definition of inductive limits), is closed because G is closed by hypothesis. Hence, in view of Proposition 3 (Chapter 2, § 5), part (*b*) of the theorem will be proved if we show that h_α is continuous for each α. In other words, in order to prove part (*b*), it is sufficient to show that a linear mapping g of a Baire l.c. space F into a generalized LF-space E with the closed graph is continuous. Evidently,

$$F = g^{-1}(E) = g^{-1}\Big(\bigcup_{n\geqslant 1} f_n(E_n)\Big) = \bigcup_{n\geqslant 1} g^{-1}(f_n(E_n)).$$

Since F is a Baire space, there exists some integer m for which $g^{-1}(f_m(E_m)) = H$ is of the second category and $\bar{H} = F$. By Lemma 5, H is a t-space. Observe $K = E_m/f_m^{-1}(0)$ is a Fréchet space (because E_m is so and $f_m^{-1}(0)$ is a closed subspace). Let h denote the $1:1$ mapping of E_m into E associated with f_m. Since the mapping $(x, y) \to (x, h(y))$ of $H \times K$ into $F \times E$ is continuous and the graph G' of $h^{-1} \circ g$ is the inverse image of the graph G of g, therefore G' is closed because G is closed by hypothesis. Hence by Theorem 5 (§ 2), $h^{-1} \circ g$ is a continuous mapping of H into K.

Now since K is complete and H is dense in F, $h^{-1} \circ g$ has a continuous extension $\tilde{g}$ which maps F into K. $h \circ \tilde{g} = g$ on H. We wish to show that this holds on F.

Suppose it is not true; then there exists $x_1 \in F$ such that

$$(x_1, h(\tilde{g}(x_1))) \notin G.$$

G being closed, it follows that there exist neighbourhoods U and V of 0 in F and E respectively such that $(x_1 + U) \times (h(\tilde{g}(x_1)) + V) \cap G = \emptyset$. Now the continuity of $h \circ g$ implies there is a neighbourhood U_1 of 0 in F such that $U_1 \subset U$ and $h(\tilde{g}(U_1)) \subset V$. But H being dense implies there exists $x_2 \in H \cap (x_1 + U_1)$, and thus

$$(x_2, g(x_2)) = (x_2, h(\tilde{g}(x_2))) \in (x_1 + U) \times (h(\tilde{g}(x_1)) + V) \cap G$$

leads to a contradiction. Therefore $h \circ \tilde{g} = g$ is continuous on F, and the proof of part (*b*) is completed.

(*a*) Since f is continuous and the quotient of a Fréchet space is also a Fréchet space, E is a generalized strict inductive limit of Fréchet spaces $E_n/f_n^{-1}(f^{-1}(0))$, where E_n and f_n are defining sequences of Fréchet spaces and of mappings respectively for E, as described in the proof of part (*b*). Therefore f can be assumed to be $1:1$ and hence f^{-1} exists. Since f is continuous, its graph is closed (Prop. 1, § 1). Moreover, the graph of f^{-1} is equal to that of f, therefore the graph of f^{-1} is closed. Hence, by part (*b*), f^{-1} is continuous. In other words, f is open.

The following theorems are particular cases of Theorem 9.

THEOREM 10. (*Dieudonné and Schwartz.*) *Let E and F be two LF-spaces. Then:*

(*a*) *A linear continuous mapping of E onto F is open.*

(*b*) *A linear mapping of F into E, the graph of which is closed in $F \times E$, is continuous.*

THEOREM 11. (*Köthe.*) *Let E and F be two generalized LF-spaces. Then:*

(*a*) *A linear continuous mapping of E onto F is open.*

(*b*) *A linear mapping of F into E with the closed graph is continuous.*

Definition 1. Let E_α $(\alpha \in A)$ be a family of Banach spaces. The inductive limit E of E_α's is called a (β)-*space* [12].

(1) (*a*) Each Fréchet space is a (β)-space. (*b*) Every generalized LF-space (in particular every LF-space) is a (β)-space. (*c*) Each bornological and quasi-complete l.c. space is a (β)-space. (*d*) The quotient of a (β)-space is a (β)-space. (*e*) Each (β)-space is a t-space. See [12].

THEOREM 12. (*Grothendieck.*) *Let E be a generalized LF-space and F a (β)-space. Then:*

(*a*) *A linear and continuous mapping of E onto F is open.*

(*b*) *A linear mapping of F into E, the graph of which is closed in $F \times E$, is continuous.*

4

B-COMPLETENESS AND THE OPEN-MAPPING THEOREM

1. Definition of B-complete spaces

In Chapter 3 we tried to find pairs E and F of l.c. spaces for which statements (A) and (B) (Chapter 3, § 1) are true. To this end we proved several theorems. Continuing our search for such pairs, an apparent generalization of Theorems 3, 7, and 9 (Chapter 3) is suggested to the reader as follows: If E is a complete l.c. space and F a t-space then statements (A) and (B) (Chapter 3, § 1) are true. Unfortunately this generalization is not true as will be shown in the sequel. One needs a stronger notion than that of completeness in order to have the open mapping and closed graph theorems. This is exactly where B-completeness comes in. The notion of B-completeness is due to V. Pták [28]. The results of this chapter are based on [28], [31] and [32]. Because of its repeated use we first prove the following:

Proposition 1. *Let E_u be an infinite-dimensional Banach space and let E_ω denote the set E with the finest locally convex topology ω. Let i denote the identity mapping: $E_\omega \to E_u$. Then:*

(*a*) *i is 1:1 and onto.*

(*b*) *i is continuous (therefore almost continuous).*

(*c*) *i is almost open.*

(*d*) *The graph of i is closed.*

(*e*) *i is* not *open.*

Proof. (*a*) is trivial. (*b*) is true because $\omega \supset u$. (*c*) is true due to (*a*) and the fact that E_u is a t-space (Chapter 2, § 6, Prop. 6). (*d*) follows from (*b*) using Proposition 1 (Chapter 3, § 1). Further, $E'_\omega = \prod_{\alpha \in H} R_\alpha$ (where R_α is the copy of the reals and H a Hamel basis of E); and E'_u is dense in E'_ω. Hence $E'_\omega \supset E'_u$ and $E'_\omega \neq E'_u$ imply that ω is strictly finer than u. This proves (*e*), because if i were open then ω would have to be equal to u.

Anticipating a result of Chapter 5 (Proposition 15), $E_\omega = \sum_{\alpha \in H} R_\alpha$ is complete. It is also easy to see that E_ω is a bornological t-space (because each R_α is so, and under the direct sum construction each of these properties is preserved). Hence Proposition 1 implies the following:

COROLLARY 1. *Let E be a complete l.c. space (and/or a t-space) and F a t-space (and/or a complete l.c. space). Then a linear continuous mapping of E onto F need* not *be open.*

Definition 1. (*a*) An l.c. space E is said to be *B-complete* if a linear continuous and almost open mapping of E onto any l.c. space F is open.

(*b*) An l.c. space E is said to be *B_r-complete* if a linear continuous almost open and one-to-one mapping of E onto any l.c. space F is open.

PROPOSITION 2. *Every B-complete space is B_r-complete.*

Proof. The proof is trivial. The converse is an open question.

PROPOSITION 3. *Every Fréchet space (in particular, a Banach space) is B-complete.*

Proof. This follows from Theorem 2 (Chapter 3, § 2).

A deeper result, namely that a B-complete l.c. space is complete, will be given later on (Chap. 5, § 3, Prop. 10). The fact that the converse is not true follows immediately from Proposition 1.

Definition 2. Let E_u be an l.c. space and E' its dual. A subspace Q of E' is said to be *almost closed* if, for each u-neighbourhood U of 0 in E, $Q \cap U^0$ is closed in the relative $\sigma(E', E)$-topology on U^0.

THEOREM 1. *Let E_u be an l.c. space and E' its dual. The following statements are equivalent:*

(*a*) *E_u is B-complete.*

(*b*) *Each almost closed subspace of E' is $\sigma(E', E)$-closed.*

Proof. (*a*) $\Rightarrow$ (*b*): Let Q be an almost closed subspace of E' and let φ denote the mapping $E_u \to E/Q^0 = F$. Let v be the l.c. topology on F defined by the collection of sets $\{\varphi(Q \cap U^0)^0\}$, where U runs over all u-neighbourhoods of 0 in E. Since for each $V = \varphi(Q \cap U^0)^0$,

$$\varphi^{-1}(V) = \varphi^{-1}(\varphi(Q \cap U^0)^0) \supset (Q \cap U^0)^0 \supset U,$$

φ is a continuous mapping of E_u onto F_v. Furthermore, for each U, $(\varphi(U))^Q = Q \cap U^0$ (because the identity mapping $Q \to E$ is weakly continuous) implies

$$v\varphi(U) = \varphi(\varphi(U))^{QE} = \varphi(Q \cap U^0)^0 = V,$$

where $(\varphi(U))^{QE}$ denotes the polar of $(\varphi(U))^Q$ in E, which itself is the polar of $\varphi(U)$ in Q, and $v\varphi(U)$ is the v-closure of $\varphi(U)$. This proves that φ is almost open. Hence by (*a*), φ is open. In other words, φ is the canonical mapping and therefore $Q = Q^{00}$ is closed due to Proposition 13 (*b*) (Chapter 2, § 9).

(*b*) $\Rightarrow$ (*a*): Let f be a linear continuous and almost open mapping of E_u onto any l.c. space F_v. Let f' denote the transpose mapping of F' into E'. Clearly f' is a homeomorphism (into). Let F' be identified with its image $f'(F') = Q$. It is easy to see that, for each neighbourhood U of 0 in E,

$$Q \cap U^0 = (f(U))^Q = (vf(U))^Q = V^Q$$

(because $vf(U) = V$, by almost openness of f) which is $\sigma(F', F)$-compact and hence $\sigma(E', E)$-closed. In other words, Q is almost closed. Hence by (*b*), Q is $\sigma(E', E)$-closed. From this it follows that f is open (Chap. 2, § 9, Prop. 19). This completes the proof.

The l.c. spaces satisfying (*b*) of Theorem 1 are known as *fully complete* spaces, due to Collins [5]. Therefore Theorem 1 says that an l.c. space is *B*-complete if and only if it is fully complete. We shall use only the term *B*-complete in the sequel.

Proposition 4. *Let E_u be a B-complete space and M its closed subspace. Then M is B-complete.*

Proof. Let Q be a subspace of the dual $M' = E'/M^0$ of M, such that $Q \cap (M \cap U)^0$ is $\sigma(M', M)$-closed for each u-neighbourhood U of 0 in E. If φ denotes the canonical mapping: $E' \to E'/M^0$, then clearly $\varphi^{-1}(Q)$ is a vector subspace of E'. Further, for each U,

$$U^0 \cap \varphi^{-1}(Q) \subset \varphi^{-1}((U \cap M)^0 \cap Q) \subset \varphi^{-1}(Q).$$

Hence $$U^0 \cap \varphi^{-1}(Q) = U^0 \cap \varphi^{-1}((U \cap M)^0 \cap Q).$$

But $(U \cap M)^0 \cap Q$ being closed in M', $\varphi^{-1}((U \cap M)^0 \cap Q)$ is $\sigma(E', E)$-closed in E' and therefore $U^0 \cap \varphi^{-1}(Q)$ is $\sigma(E', E)$-closed. Since E is *B*-complete, $\varphi^{-1}(Q)$ is closed by Theorem 1. Hence Q is $\sigma(M', M)$-closed in M'.

Proposition 5. *Let E_u and F_v be two l.c. spaces. Let f be a linear continuous and almost open mapping of E_u onto F_v. If E_u is B-complete then F_v is also B-complete.*

Proof. Let Q be a subspace of F'_v such that $Q \cap V^0$ is $\sigma(F', F)$-closed for each v-neighbourhood V of 0 in F. Clearly $f'(Q)$ is a subspace of E', where f' denotes the transpose mapping: $F' \to E'$ which is a homeomorphism into. Since

$$f'(Q) \cap U^0 = f'((f(U))^0 \cap Q)$$

for each u-neighbourhood U of 0 in E, $f'((f(U))^0 \cap Q)$ is $\sigma(E', E)$-compact because $(f(U))^0 \cap Q = V^0 \cap Q$ is a $\sigma(F', F)$-closed (by assumption) subset of a $\sigma(F', F)$-compact set V^0, and f' is weakly continuous. Therefore $f'(Q) \cap U^0$ being $\sigma(E', E)$-closed for each u-neighbourhood U of 0 in E implies $f'(Q)$ is $\sigma(E', E)$-closed due to Theorem 1, since E is B-complete by hypothesis. Hence Q is $\sigma(F', F)$-closed in F'. This completes the proof.

COROLLARY 2. *Let E be a B-complete l.c. space and M a closed subspace of E. Then E/M is B-complete.*

Proof. This is a particular case of Proposition 5, since the canonical mapping f: $E \to E/M$ is linear, continuous, and open.

(1) An arbitrary product of B-complete spaces is *not* B-complete. For, if so, then an arbitrary product of Banach spaces would be B-complete, since each Banach space is B-complete (Prop. 3). But then every complete l.c. space, being a closed subspace of a product of Banach spaces (Chapter 2, § 4, (5) (*d*)), must be B-complete due to Proposition 4. But this is false due to the example in Proposition 1.

(2) (*a*) An arbitrary direct sum of B-complete spaces is *not* B-complete, because in Proposition 1, $E_\omega = \sum_{\alpha \in A} R_\alpha$ is not B-complete due to Corollary 1, while each R_α, being a copy of the reals, is B-complete. (*b*) Even a countable direct sum of B-complete spaces is *not* B-complete. This follows from an example due to A. Grothendieck [11]. (*c*) It is *not* known if finite direct sums or direct products of B-complete spaces are B-complete.

PROPOSITION 6. *Let E_u be a B-complete l.c. space and let u_1 be an l.c. topology on E such that $u_1 \supset u$ and $u_1 \sim u$. Then E_{u_1} is B-complete.*

Proof. Since $E'_{u_1} = E'_u$, the proposition follows immediately from Theorem 1.

PROPOSITION 7. *Let E_u be a Fréchet space and E' its dual. Then E' is B-complete for any locally convex topology v such that $\tau(E', E) \supset v \supset c$* (see Chapter 2, § 9).

Proof. In view of Proposition 6, it is sufficient to show that E'^c is B-complete. Let Q be a subspace of E such that $Q \cap C^{00}$ is $\sigma(E, E')$-closed and hence u-closed (because $u \supset \sigma(E, E')$) for each u-compact set C of E. Let $x \in \bar{Q}$ and let x_n $(n \geqslant 1)$ be a sequence in Q such that $x_n \to x$. The sequence x_n $(n \geqslant 1)$, being convergent, is precompact, and therefore the set $P = \{x_n \ (n \geqslant 1)\}$ is relatively compact (because E_u is complete). Hence P^0 is a c-neighbourhood of 0 in E' and P^{00} is compact. Since $x_n \in Q \cap P^{00}$ for each n and $x_n \to x$, therefore $x \in Q \cap P^{00}$ because the latter is closed by assumption. Hence $x \in Q$. This shows that Q is closed.

2. The $v(u)$-topology

Let u and v be two topologies on a vector space E such that E_u and E_v are TVS's. Let $v(u)$ denote the topology whose fundamental system of neighbourhoods of 0 consists of u-neighbourhoods of 0 which are v-closed. In other words, if U is a u-closed neighbourhood of 0, then vU (i.e. the v-closure of U) is a member of a basis of $v(u)$-neighbourhoods of 0. Clearly $v(u)$ is compatible with the vector space structure of E, i.e. $E_{v(u)}$ is a topological vector space. Moreover, for each U, $vU \supset U$ implies that $u \supset v(u)$. If u is locally convex, so is $v(u)$.

The main feature of the $v(u)$-topology is that it need *not* be Hausdorff even though u and v both are Hausdorff. The observation that $v(u)$ is Hausdorff if the identity mapping $E_u \to E_v$ has the closed graph, is due to A. Robertson and W. Robertson [31]. The notation $v(u)$ is due to V. Pták [29].

THEOREM 2. *Let u and v be two locally convex Hausdorff topologies on a vector space E. The following statements are equivalent:*

(*a*) *$v(u)$ is Hausdorff.*

(*b*) *For any two distinct points x_1 and x_2 of E there exist a u-neighbourhood U and a v-neighbourhood V of 0 such that $(x_1+U) \cap (x_2+V) = \emptyset$.*

(*c*) *The set $\Delta = \{(x,x);\, x \in E\}$ is closed in $E_u \times E_v$.*

(*d*) *The set H of v-continuous functionals on E is dense in E'_u.*

Proof. The pattern of the proof is: (*a*) $\Rightarrow$ (*b*) $\Rightarrow$ (*c*) $\Rightarrow$ (*d*) $\Rightarrow$ (*a*).

For (*a*) $\Rightarrow$ (*b*), let $x_1 \neq x_2$. By (*a*) there exists a u-neighbourhood U of 0 such that $x_2 \notin x_1+vU = v(x_1+U)$. This shows that there exists a v-neighbourhood V of 0 such that (*b*) is satisfied.

For (*b*) $\Rightarrow$ (*c*), let $(x,y) \notin \Delta$. By (*b*) there exist u- and v-neighbourhoods U and V of 0 such that $(x+U) \cap (y+V) = \emptyset$. That means $(x+U)\times(y+V)$ is a neighbourhood of (x,y) which does not meet Δ. This establishes (*c*).

For (*c*) $\Rightarrow$ (*d*), suppose $x_0 \neq 0$ such that $\langle x_0, H\rangle = 0$. Since the diagonal set Δ is a closed subspace, there exists, by the Hahn–Banach Theorem, a closed hyperplane, or in other words a continuous functional $z' = (x',y')$ on $E_u \times E_v$ such that

$$z'(\Delta) = 0 \quad \text{and} \quad z'(x_0, 0) \neq 0,$$

where $x' \in E'_u$ and $y' \in E'_v$. But $z'(\Delta) = 0$ implies $z'(x,x) = 0$ for each

$(x,x) \in \Delta$. Thus $o = z'(x,x) = \langle x,x'\rangle + \langle x,y'\rangle$. Since y' is v-continuous on E, x' is also v-continuous on E. Therefore $x' \in H$. But then $\langle x_0,x'\rangle = \langle x_0,x'\rangle + \langle 0,y'\rangle = z'(x_0,0) = o$ is a contradiction. Hence $x_0 = 0$, and (d) is established.

For (d) $\Rightarrow$ (a), let $x_0 \in vU$ for all U. Suppose $x_0 \neq 0$. There exists an $x' \in H$ such that $\langle x_0,x'\rangle = \alpha > o$. The set U of all $x \in E$ such that

$$\langle x,x'\rangle \leqslant \tfrac{1}{2}\alpha$$

is clearly a u-closed u-neighbourhood of 0. Since, by definition, x' is v-continuous, U is a v-closed u-neighbourhood of 0. This shows $x_0 \in vU = U$ for all U which is a contradiction, since u is Hausdorff. Hence $x_0 = 0$. Thus the theorem is completely established.

The notation $v(u)$ is usefully applied in the following proposition which is another version of Proposition 8 ([3], Chapitre 1, § 1, N° 5).

Proposition 8. *Let u and v be two locally convex topologies on a vector space E such that $u \supset v$. If E_v is complete, so is $E_{v(u)}$.*

Proof. Since $u \supset v$ (i.e. the identity mapping $i\colon E_u \to E_v$ is continuous and therefore has the closed graph), $v(u)$ is Hausdorff by Theorem 2. Further, $u \supset v(u)$ (by definition) and $v(u) \supset v$ (due to $u \supset v$) imply $u \supset v(u) \supset v$. Let $\mathscr{K}$ be a Cauchy filter in $E_{v(u)}$. From $v(u) \supset v$ it follows that $\mathscr{K}$ is a Cauchy filter in E_v and therefore converges to x_0 by hypothesis. Let V be a circled $v(u)$-closed neighbourhood of 0. By assumption there exists a set A in $\mathscr{K}$ such that $A - A \subset V$, i.e. $A \subset x_1 + V$ if $x_1 \in A$. Since $x_0 \in vA \subset v(x_1+V) = x_1+V$, therefore $x_1 \in x_0+V$. This shows that x_0 is the limit point of $\mathscr{K}$ in the topology $v(u)$. This completes the proof.

Proposition 9. *Let E_u be an l.c. space and v another locally convex topology on E such that $u \supset v$. Then the following statements are equivalent:*

(a) *The identity mapping $E_u \to E_v$ is almost open.*

(b) $v(u) = v$.

Proof. (a) $\Rightarrow$ (b): Since $u \supset v$, $u \supset v(u) \supset v$. Now (a) implies that for each u-neighbourhood U of 0, vU is a v-neighbourhood. In other words, $v(u) \subset v$. Combining this with the earlier inequality we have $v(u) = v$. The other implication, i.e. (b) $\Rightarrow$ (a), is trivial.

Proposition 10. *Let E be an l.c. space under two topologies u and v such that $u \supset v$. If E_u is a t-space and $v(u) = v$, then E_v is a t-space.*

Proof. In view of Proposition 9, this proposition is a particular case of Proposition 5 (Chapter 2, § 6).

Definition 3. Let u and v be two l.c. topologies on a vector space E. The $v(u)$-topology is always coarser than u. However, if $v(u) = u$, then it is said that the *closed neighbourhood condition* holds.

(1) (*a*) It is easy to see that if $v \supset u$ then $v(u) = u$. (*b*) A necessary and sufficient condition for the closed neighbourhood condition to hold is that $v(u) \supset u$.

PROPOSITION 11. *An l.c. space E_v is a t-space if and only if, for each l.c. topology u for which E_u is an l.c. space, $v(u) = u$ implies $v \supset u$.*

Proof. Suppose E_v is a t-space and let u be an l.c. topology such that $v(u) = u$. Let U be a convex circled u-closed u-neighbourhood of 0. vU, being convex, circled, absorbing, and v-closed, is a v-neighbourhood of 0 by supposition. In other words, there exists a v-neighbourhood V of 0 such that $vU \supset V$. But $v(u) = u$ implies that $vU = U \supset V$ and hence $v \supset u$.

Conversely, let u denote the topology for which v-barrels form a fundamental system of neighbourhoods of 0. u is clearly an l.c. topology. Since each member of the fundamental system is v-closed, $v(u) = u$. But then the assumption implies that $v \supset u$. In other words, each v-barrel, being a u-neighbourhood, is a v-neighbourhood of 0. This proves the proposition.

Definition 4. Let E and F be two l.c. spaces and f a linear mapping of E into F. It is said that the *filter condition holds with respect to f* if, for each Cauchy filter $\mathscr{K}$ in E,

$$f(\mathscr{K}) \to f(x_0) \quad \text{implies} \quad \mathscr{K} \to x_0.$$

(For some other variations of this definition see [32].)

PROPOSITION 12. *Let u and v be two l.c. topologies on a vector space E such that $u \supset v$ and $v(u) = u$. Then the filter condition holds with respect to the identity mapping i: $E_u \to E_v$.*

Proof. Let $\mathscr{K}$ be a Cauchy filter in the u-topology such that $i(\mathscr{K}) = \mathscr{K} \to x_0$ in the v-topology. Since $v(u) = u$, by the same argument as used in Proposition 8, $\mathscr{K} \to x_0$ in the u-topology. This proves the proposition.

3. Ultra-barrelled spaces

Definition 5. A TVS E_v is said to be *ultra-barrelled* if for each linear topology u (i.e. a topology u in which E is a TVS), $v(u) = u$ implies $v \supset u$.

(1) (*a*) One observes that ultra-barrelled spaces need not be locally convex. (*b*) Every locally convex ultra-barrelled space is barrelled

(Prop. 11). The converse is not true [32]. (*c*) Each Baire TVS (in particular, each F-space) is ultra-barrelled. (*d*) As in the case of barrelled spaces, the quotient and the completion of ultra-barrelled spaces are ultra-barrelled. (*e*) From Proposition 11 it follows that Proposition 6 of Chapter 2 is true if one replaces t-spaces by ultra-barrelled spaces. (*f*) Hence Theorems 3 and 5 of Chapter 3 remain valid if we replace the t-spaces by ultra-barrelled spaces and Fréchet spaces by F-spaces.

The notion of ultra-barrelled spaces and the results of this section are due to W. Robertson [32].

Proposition 13. *Let E_v be ultra-barrelled and F a linear subspace of $\hat{E}_v$ such that $E_v \subset F \subset \hat{E}_v$. Then F is ultra-barrelled.*

Proof. $\hat{E}$ is ultra-barrelled as remarked above. Let u denote a linear topology on F such that $v(u) = u$. Here v is understood to be the relativization of the extension of v from E to $\hat{E}$. (That is, for each v-neighbourhood V of 0 in E, $\hat{V}$ is a neighbourhood of 0 in $\hat{E}$ and therefore $\hat{V} \cap F$ is a v-neighbourhood of 0 in F.) Clearly u induces a topology u_1 on E such that $v(u_1) = u_1$. For each v-closed u-neighbourhood U of 0 in F, $U \cap E$ is a v-neighbourhood of 0 in E because E_v is ultra-barrelled. Hence $(\widehat{U \cap E})$ is a v-neighbourhood of 0 in $\hat{E}$, and therefore $\hat{U} \supset (\widehat{U \cap E})$ is a v-neighbourhood of 0 in $\hat{E}$. But F is dense in $\hat{E}$ because E is so. Hence $\hat{U} \cap F = vU = U$ is a v-neighbourhood of 0 in F. This shows that $v \supset u$. In other words, F is ultra-barrelled.

Lemma 1. *Let E and F be two TVS's and f a linear and continuous mapping of E into F. Let $\hat{f}$ denote the continuous extension of f, which maps $\hat{E}$ into $\hat{F}$. Then $f^{-1}(0) = \hat{f}^{-1}(0)$ if and only if the filter condition holds.*

Proof. Suppose the filter condition holds. Clearly $f^{-1}(0) \subset \hat{f}^{-1}(0)$ because $f(x) = \hat{f}(x)$ for $x \in E$. To show that $f^{-1}(0) \supset \hat{f}^{-1}(0)$, let $\hat{x} \in \hat{E}$ such that $\hat{f}(\hat{x}) = 0$. Let $\mathscr{K}$ be a Cauchy filter on E which is a base of a convergent filter $\mathscr{K}'$ which converges to $\hat{x}$. By continuity of $\hat{f}$, $\hat{f}(\mathscr{K}') \to \hat{f}(\hat{x}) = 0$. Consider the sets $f(\mathscr{K})$ and $\hat{f}(\mathscr{K}')$ on $f(E)$. For each K' in $\mathscr{K}'$, there exists a subset K of K' such that K is in $\mathscr{K}$ and $f(K) = \hat{f}(K) \subset \hat{f}(K') \cap f(E)$. Hence $f(\mathscr{K})$ is finer than the trace of $\hat{f}(\mathscr{K}')$ on $f(E)$. Hence $f(\mathscr{K}) \to 0$. By the filter condition it follows that $\mathscr{K}$ converges to a point in E. Hence $\hat{x} \in E$ implies $f(\hat{x}) = 0$. In other words, $\hat{x} \in f^{-1}(0)$.

For the converse, let $\mathscr{K}$ be a Cauchy filter on E such that $f(\mathscr{K}) \to f(x)$. Then there is a point $\hat{x} \in \hat{E}$ and a filter $\mathscr{K}_1$ on $\hat{E}$ with a base $\mathscr{K}$ such that $\mathscr{K}_1 \to \hat{x}$. Hence $\hat{f}(\mathscr{K}_1) \to \hat{f}(\hat{x})$. By the same argument as above, $f(\mathscr{K})$ is coarser than the trace of $\hat{f}(\mathscr{K}_1)$ on $f(E)$. But $f(\mathscr{K}) \to f(x)$ and therefore $\hat{f}(\hat{x}) = f(x)$. Now if $f^{-1}(0) = \hat{f}^{-1}(0)$, then $x - \hat{x} \in f^{-1}(0)$ and so $\hat{x} \in E$. Therefore $\mathscr{K}$ converges to $x \in E$.

THEOREM 3. (*Robertson* [32].) *Let E be a metrizable TVS and F an ultra-barrelled TVS. Let f be a linear and continuous mapping of E onto F. Then f is open if the filter condition holds.*

Proof. Replacing E and F by their completions $\hat{E}$ and $\hat{F}$ respectively, we observe that the continuous extension $\hat{f}$ (of f) which maps an F-space $\hat{E}$ onto an ultra-barrelled space $\hat{f}(\hat{E})$ is open due to ((1) (f)) (the fact that $\hat{f}(\hat{E})$ is ultra-barrelled follows from Proposition 13). Let U be a closed neighbourhood of 0 in E. Then $\hat{f}(\hat{U})$ is a neighbourhood of 0 in $\hat{f}(\hat{E})$ and therefore $\hat{f}(\hat{U}) \cap F$ is a neighbourhood of 0 in F. We show that $\hat{f}(\hat{U}) \cap F \subset f(U)$. Let $y \in \hat{f}(\hat{U}) \cap F$. Then there exists $\hat{x} \in \hat{U}$ such that $\hat{f}(\hat{x}) = y$, and $x \in E$ such that $f(x) = y$, because f is onto. Therefore $\hat{f}(\hat{x}) - f(x) = \hat{f}(\hat{x} - x) = 0$. By Lemma 1, it follows that $\hat{x} = x \in E$. Thus $\hat{x} \in \hat{U} \cap E = U$ (because U is closed in E) shows that

$$y = \hat{f}(\hat{x}) \in \hat{f}(U) = f(U).$$

This proves the theorem.

Definition 6. Let g be a linear mapping of a TVS F into a TVS E. It is said that the *inverse filter condition holds* if for a convergent filter base $\mathscr{K}$ on F such that $g(\mathscr{K})$ is Cauchy, it follows that $g(\mathscr{K})$ is convergent to a point in $g(F)$.

THEOREM 4. (*Robertson* [32].) *Let F be an ultra-barrelled TVS and E a metrizable TVS. Let g be a linear mapping of F into E, the graph of which is closed in $F \times E$. Then g is continuous provided the inverse filter condition holds.*

Proof. Note that g can be regarded as a mapping of F into $\hat{E}$. Let G be the graph of g in $F \times E$. We show that G is closed in $F \times \hat{E}$. Let $(x, y) \in \bar{G}$, then $(x+U) \times (y+V)$ meets G for each U and V, neighbourhoods of 0 in F and $\hat{E}$ respectively. That means

$$A_{UV} = (x+U) \cap g^{-1}(y+V) \neq \emptyset.$$

Let $\mathscr{K}$ be the filter generated by the sets A_{UV}, where U and V run over a fundamental system of neighbourhoods of 0 in F and $\hat{E}$ respectively. Then $\mathscr{K} \to x$ and $g(\mathscr{K}) \to y$. Therefore $g(\mathscr{K})$ is Cauchy and so, by the inverse filter condition, $g(\mathscr{K})$ converges to a point in $g(F)$. In other words, $y \in g(F) \subset E$. This shows that $(x, y) \in \bar{G} \cap (F \times E)$. But G being closed in $F \times E$, $(x, y) \in G$. Therefore G is closed in $F \times \hat{E}$. Hence by (1)(*f*), g is a continuous mapping of F into $\hat{E}$. Since the mapping $E \to \hat{E}$ is continuous, g is continuous.

4. B_r-complete l.c. spaces

By Proposition 2, every B-complete space is B_r-complete. In view of Theorem 1, we should like to have a similar characterization for B_r-complete spaces. First we need the following proposition, which is due to V. Pták [29], as are other results of this section.

PROPOSITION 14. *Let E_u be an l.c. space and E' its dual. Let v be another l.c. topology on E such that $u \supset v$. Let Q denote the subspace of E' where $Q = E'_v$. Then the following statements are equivalent:*

(*a*) $v(u) \sim v$ (for $\sim$ see Chapter 2, § 9, Def. 14).

(*b*) *For each u-neighbourhood U of 0 in E, $Q \cap U^0$ is $\sigma(E'_u, E_u)$-closed in E'_u.*

Proof. (*a*) ⇒ (*b*). For each u-neighbourhood U of 0, vU being a $v(u)$-neighbourhood of 0, $(vU)^Q$ is $\sigma(E'_{v(u)}, E_{v(u)})$-compact. But then (*a*) implies that $(vU)^Q$ is $\sigma(E'_v, E_v)$-compact. Since $u \supset v$ implies the identity mapping: $E'_v \to E'_u$ is a homeomorphism into, therefore $(vU)^Q$ is $\sigma(E'_u, E_u)$-compact and hence $\sigma(E'_u, E_u)$-closed. Now the equation

$$(vU)^Q = U^Q = Q \cap U^0$$

proves (*b*).

(*b*) ⇒ (*a*): Suppose $Q \cap U^0$ is $\sigma(E'_u, E_u)$-closed for each U. Since U^0 is $\sigma(E'_u, E_u)$-compact, it follows that $Q \cap U^0$ is $\sigma(E'_u, E_u)$-compact and therefore $\sigma(E'_v, E_v)$-compact because of the homeomorphism (into) of the mapping: $E'_v \to E'_u$. One can assume without loss of generality that each U is circled. Then each $Q \cap U^0$ being circled, convex, and $\sigma(E'_v, E_v)$-compact such that $\bigcup (Q \cap U^0) = Q$, the $\mathfrak{S}$-topology v' on E, where $\mathfrak{S}$ consists of sets $\{Q \cap U^0\}$, is compatible with the duality between Q and E_v, i.e. $v' \sim v$ by Theorem 6 (Chapter 2, § 9). But since

$$(Q \cap U^0)^0 = U^{QE} = vU$$

for each U, $v' = v(u)$. This completes the proof.

THEOREM 5. *Let E_u be an l.c. space and E' its dual. The following statements are equivalent:*

(*a*) *E_u is B_r-complete.*

(*b*) *For each l.c. topology v on E, $u \supset v$ and $v(u) \sim v$ imply $u \sim v$.*

(*c*) *For each l.c. topology v on E, $u \supset v$ and $v(u) = v$ imply $v = u$.*

(*d*) *For each l.c. topology v on E, $u \supset v$ and $v(u) \subset v$ imply $v = u$.*

(*e*) *Each almost closed dense subspace Q of E' coincides with E'.*

Proof. Since $u \supset v$ always implies $u \supset v(u) \supset v$, the conditions $v(u) = v$ and $v(u) \subset v$ are equivalent. Hence $(c) \Leftrightarrow (d)$.

Further let f be a linear, continuous, 1:1 and almost open mapping of E_u onto an l.c. space F. Because of one-to-oneness one can always replace F by E itself with an l.c. topology v, and f by the identity mapping. Thus in view of Proposition 9 (§ 2), $(a) \Leftrightarrow (c)$.

$(b) \Rightarrow (c)$: Let v be an l.c. topology on E such that $u \supset v$ and $v(u) = v$. Therefore, by (b), $u \sim v$ because $v(u) = v$ implies $v(u) \sim v$. For each u-closed convex neighbourhood U of 0, vU is a $v(u)$-neighbourhood of 0. But

$$U = U^{00} \supset vU$$

(because $u \sim v$) shows $u \subset v(u)$. But in general $u \supset v(u)$ proves $v(u) = u$. But then $v(u) = v$ (by hypothesis) establishes (c).

$(c) \Rightarrow (e)$: Let Q be a dense subspace of E' such that $Q \cap U^0$ is σ-closed in E' for each convex, circled, u-neighbourhood U of 0. Clearly $Q' = E$ because Q is dense in E' and σ is the induced topology $\sigma(E', E)$ on Q. Let v denote the $\mathfrak{S}$-topology on E, where $\mathfrak{S}$ consists of sets $(Q \cap U^0)$. Since U^0 is $\sigma(E', E)$-compact, $Q \cap U^0$, being closed, is compact and also convex and circled. Therefore by Theorem 6 (Chapter 2, § 9), v is compatible with the duality between E_v and Q. In other words, $E'_v = Q$. Since for each U, $Q \cap U^0 \subset U^0$ implies $(Q \cap U^0)^0 \supset U^{00} \supset U$, therefore $u \supset v$. Further, for each U,

$$vU = (U^Q)^0 = (Q \cap U^0)^0$$

shows that $v(u) = v$. Hence by (c), $u \supset v$ and $v(u) = v$ imply $u = v$. In other words, $Q = E'_v = E'_u$. This proves (e).

$(e) \Rightarrow (b)$: Let v be an l.c. topology on E such that $u \supset v$ and $v(u) \sim v$. Let $Q = E'_v$. Then by Proposition 14, for each u-neighbourhood U of 0, $Q \cap U^0$ is $\sigma(E', E)$-closed. Clearly Q is dense in E'_u. Hence by (e), $Q = E'_u$. This proves that $u \sim v$.

The following characterization of B_r-complete spaces is also useful.

Proposition 15. *Let E_u be an l.c. space. Then the following statements are equivalent:*

(a) E_u is B_r-complete.

(b) For each locally convex topology v on E, if $v(u)$ is Hausdorff and $v(u) \subset v$, then $v \supset u$.

(c) For each l.c. space F, a 1:1 linear mapping f of E onto F, the graph of which is closed in $E \times F$, is open provided f is almost open.

Proof. To consider a 1:1 linear mapping f of E_u onto F is equivalent to considering another l.c. topology v on E. By Theorem 2, to say that the graph of the mapping: $E_u \to E_v$ is closed is equivalent to saying that $v(u)$ is Hausdorff. Trivially, f being almost open is equivalent to $v(u) \subset v$. Therefore (*b*) and (*c*) are equivalent.]

(*a*) $\Rightarrow$ (*b*): Let $v(u) = w$. Then $w \subset v$ by hypothesis, and $w = v(u) \subset u$ always. Since for each u-neighbourhood U of 0, $vU \subset wU$, therefore $v(u) \supset w(u)$. Since E_u is B_r-complete, by part (*d*) of Theorem 5, $u \supset w$ and $w(u) \subset w$ imply $u = w$. But $w = v(u) \subset v$ establishes (*b*). The implication (*b*) $\Rightarrow$ (*a*) is trivial.

(1) (*a*) A closed subspace of a B_r-complete space is B_r-complete. (*b*) If E is B_r-complete and M a closed subspace of E, then E/M is B_r-complete.

5. The open-mapping and closed-graph theorems for B_r-complete spaces

Before we prove the open mapping and closed graph theorems for B_r-complete spaces, we need a strengthened form of condition (*d*) of Theorem 2 as follows:

LEMMA 2. *Let f be a linear mapping of an l.c. space E_u into another l.c. space F_v. The following statements are equivalent:*

(*a*) *The graph G of f is closed in $E \times F$.*

(*b*) *The subspace H of F' consisting of those $y' \in F'$ for which $\langle f(x), y'\rangle$ is continuous on E_u, is dense in F'.*

Proof. (*a*) $\Rightarrow$ (*b*): Let $y_0 \in F$ such that $\langle y_0, H\rangle = 0$. In order to show that H is dense we show that $y_0 = 0$. For this it is sufficient to show that $z_0 = (0, y_0) \in G$. Suppose $z_0 \notin G$. Since G is closed, by the Hahn–Banach theorem there exists an element z' in $E' \times F'$ such that

$$\langle G, z'\rangle = 0 \quad \text{and} \quad \langle z_0, z'\rangle \neq 0.$$

But every $z' = (x', y')$ where $x' \in E'$ and $y' \in F'$, therefore the first equation implies

$$\langle x, x'\rangle + \langle f(x), y'\rangle = 0.$$

Since $\langle x, x'\rangle$ is continuous for each $x \in E$, $\langle f(x), y'\rangle$ is also continuous. Therefore $y' \in H$. But then $\langle 0, x'\rangle + \langle y_0, y'\rangle = \langle z_0, z'\rangle = 0$ which is a contradiction. Therefore $(0, y_0) \in G$.

(*b*) $\Rightarrow$ (*a*): Let $z_0 = (x_0, y_0)$ be any point in $E \times F$ which does not belong to G, i.e. $f(x_0) \neq y_0$. Since H is dense in F', there exists $y' \in H$ such that $\langle y_0 - f(x_0), y'\rangle \neq 0$. But then there exists $x' \in E'$ such that $\langle x, x'\rangle + \langle f(x), y'\rangle = 0$ for each $x \in E$. Hence $\langle G, z'\rangle = 0$ where

$z' = (x', y')$, and

$$\langle z_0, z'\rangle = \langle x_0, x'\rangle + \langle y_0, y'\rangle$$
$$= \langle x_0, x'\rangle + \langle f(x_0), y'\rangle + \langle y_0 - f(x_0), y'\rangle \neq 0.$$

Therefore $z_0 \notin \bar{G}$. In other words, G is closed.

LEMMA 3. *Let f be a linear mapping of an l.c. space F_v into another l.c. space E_u. Let H denote the subspace of E' consisting of all those $x' \in E'_u$ for which $\langle f(y), x'\rangle$ is continuous on F_v. If f is almost continuous then H is almost closed.*

Proof. For each v-neighbourhood V of 0 in F, $(f(V))^0 \subset H$. For, if $x' \in (f(V))^0$, the functional $\langle f(y), x'\rangle$, being bounded on a neighbourhood V of 0, is continuous and therefore $x' \in H$. Furthermore, since f is almost continuous, for each u-neighbourhood U of 0 in E_u, $\overline{f^{-1}(U)}$ is a v-neighbourhood of 0 in F_v. Putting $\overline{f^{-1}(U)} = V$, we have $(f(V))^0 \subset H$. But since

$$f(V) = f(\overline{f^{-1}(U)}) \supset U,$$

therefore $$(f(V))^0 \subset U^0.$$

Whence $$(f(V))^0 \subset H \cap U^0.$$

On the other hand, if $x' \in H \cap U^0$ then $\langle f(y), x'\rangle$ is continuous on F_v (because $x' \in H$), and $\langle x, x'\rangle \leqslant 1$ for all $x \in U$ (because $x' \in U^0$). Hence

$$\langle f(f^{-1}(U)), x'\rangle \leqslant 1$$

and by continuity of the linear functional we have

$$\langle f(\overline{f^{-1}(U)}), x'\rangle = \langle f(V), x'\rangle \leqslant 1.$$

This shows that $x' \in (f(V))^0$ and hence

$$H \cap U^0 \subset (f(V))^0 \subset H \cap U^0.$$

In other words, $H \cap U^0 = (f(V))^0$ which is closed in E' for each U. This establishes the lemma.

THEOREM 6. (*Pták* [29].) *Let F_v be an l.c. space and E_u a B_r-complete l.c. space. Let f be a linear mapping of F_v into E_u, the graph of which is closed in $F \times E$. If f is almost continuous then f is continuous.*

Proof. In view of Proposition 18 (Chapter 2, § 9), it is sufficient to show that f is continuous when F is endowed with the $\sigma(F, F')$-topology. Let H denote the set of all those $x' \in E'_u$ for which $\langle f(y), x'\rangle$ is continuous for $y \in F$. By Lemma 2, H is a dense subspace of E' (because the graph of f is closed), and by Lemma 3, it is almost closed (because f, in addition, is almost continuous). Since E_u is B_r-complete, by part (*e*) of Theorem 5, $H = E'$. This proves the theorem.

COROLLARY 3. *Let F_v be a t-space and E_u a B_r-complete l.c. space. Let f be a linear mapping of F_v into E_u, the graph of which is closed in $F \times E$. Then f is continuous.*

Proof. It is immediate from Theorem 6, by observing that F_v being a t-space implies f is almost continuous (Chapter 2, § 6, Prop. 6).

THEOREM 7. *Let E_u be a B_r-complete l.c. space and F_v an l.c. space. Let f be a linear mapping of E_u onto F_v such that the graph of f is closed in $E \times F$. If f is almost open then f is open.*

Proof. In view of Lemma 3 (Chapter 3, § 2) and ((1) (*b*)) (§ 4), it can be assumed, in addition, that f is one-to-one. Thus the inverse mapping $g = f^{-1}$ is well-defined, and has the closed graph because f does so. Almost openness of f implies almost continuity of g. Hence, by Theorem 6, g is continuous and therefore f is open.

COROLLARY 4. *Let E_u be a B_r-complete l.c. space and F_v a t-space. Let f be a linear mapping of E_u onto F_v such that the graph of f is closed in $E \times F$. Then f is open.*

Proof. By Proposition 6 (Chapter 2, § 6), f is almost open. Hence, by Theorem 7, f is open.

It is possible to strengthen Theorem 7 by assuming f to be *into* instead of *onto*. For details see Pták's paper [29].

A particular case of Theorems 6 and 7 is the following:

THEOREM 8. *Let E be a B-complete l.c. space and F a t-space. Then:*

(*a*) (*Pták* [28].) *A linear and continuous mapping f of E onto F is open.*

(*b*) (*Robertson, A. P. and W.* [31].) *A linear mapping g of F into E, the graph of which is closed in $F \times E$, is continuous.*

Replacing the generalized LF-space by a generalized strict inductive limit of B-complete l.c. spaces in Theorem 9 (Chapter 3, § 3), in view of Theorem 8 one gets a theorem due to A. P. Robertson and W. Robertson [31], namely:

THEOREM 9. *Let E be a generalized strict inductive limit of B-complete l.c. spaces and F an inductive limit of Baire l.c. spaces. Then:*

(*a*) *A linear and continuous mapping f of E onto F is open.*

(*b*) *A linear mapping g of F into E, the graph of which is closed in $F \times E$, is continuous.*

5

THE ew^*-TOPOLOGY AND VARIOUS NOTIONS OF COMPLETENESS

1. The k-extension of a topology in topological spaces

In view of the characterization of a B-complete l.c. space (Chapter 4, § 1, Theorem 1), it is apparent that the study of the k-extension (see below) of the $\sigma(E', E)$-topology is desirable.

Definition 1. (*a*) Let X_u be a Hausdorff topological space. Let $\mathscr{C}$ denote a collection of u-compact subsets of X. The *k-extension* $k(u, \mathscr{C})$ of the topology u is defined as follows: A set V is $k(u, \mathscr{C})$-open if and only if $V \cap C$ is open in the relative u-topology of C for each C in $\mathscr{C}$. Equivalently a set B is $k(u, \mathscr{C})$-closed if and only if $B \cap C$ is compact in the relative topology of C for each C in $\mathscr{C}$.

Clearly $k(u, \mathscr{C}) \supset u$. Actually $k(u, \mathscr{C})$ is the *finest* topology which coincides with u on each C in $\mathscr{C}$ and therefore the inclusion $k(u, \mathscr{C}) \supset u$ is strict in general.

(*b*) A topological space X_u is said to be a *k-space* if $k(u, \mathscr{C}) = u$, where $\mathscr{C}$ is the collection of all u-compact subsets of X.

Proposition 1. *A metrizable space X_u is a k-space.*

Proof. It is sufficient to show that $k(u, \mathscr{C}) \subset u$. Let B be a non-closed subset of X in the topology u, and let x be a limit point of B such that $x \notin B$. Let x_n ($n \geqslant 1$) be a sequence in B such that $x_n \to x$. Clearly $\{(x_n), x\} = C$ is a compact subset of X and $x \notin C \cap B$, because $x \notin B$. Therefore B is non-closed in the topology $k(u, \mathscr{C})$. This proves the proposition.

Proposition 2. *Let $\mathscr{C}_1$ and $\mathscr{C}_2$ be two classes of u-compact subsets of a Hausdorff topological space X_u such that each C_1 in $\mathscr{C}_1$ is also in $\mathscr{C}_2$. Then*

$$k(u, \mathscr{C}_1) \supset k(u, \mathscr{C}_2).$$

If $\mathscr{C}_1 = \mathscr{C}_2$, then $k(u, \mathscr{C}_1) = k(u, \mathscr{C}_2)$.

Proof. Let W be a $k(u, \mathscr{C}_2)$-open set, i.e. $W \cap C_2$ is open for each C_2 in $\mathscr{C}_2$. But since each C_1 in $\mathscr{C}_1$ is also in $\mathscr{C}_2$ by hypothesis, therefore $W \cap C_1$ is open for each C_1 in $\mathscr{C}_1$, i.e. W is $k(u, \mathscr{C}_1)$-open.

The second part follows from the fact that '$\mathscr{C}_1 = \mathscr{C}_2$' is equivalent to '$\mathscr{C}_1 \subset \mathscr{C}_2$ and $\mathscr{C}_2 \subset \mathscr{C}_1$'.

PROPOSITION 3. *Let X_u be a Hausdorff topological space and $\mathscr{C}$ the collection of all its u-compact subsets. Then X, endowed with the $k(u,\mathscr{C})$-topology, is a k-space.*

Proof. In view of Proposition 2, it is sufficient to show that $\mathscr{C} = \mathscr{C}_1$, where $\mathscr{C}_1$ is the collection of all $k(u,\mathscr{C})$-compact sets. Clearly $\mathscr{C}_1 \subset \mathscr{C}$ because $k(u,\mathscr{C}) \supset u$. On the other hand, each u-compact C is $k(u,\mathscr{C})$-compact because, on each C in $\mathscr{C}$, the $k(u,\mathscr{C})$-topology induces the same topology as that induced by u. Therefore $\mathscr{C}_1 = \mathscr{C}$.

2. The *k*-extension of a topology in a topological vector space

We should like to make use of the above results in the case when X_u is a topological vector space. Unfortunately, the trouble in this case is that the k-extension of a topology u under which X is a TVS is not compatible with the vector space structure. More precisely, if X_u is a TVS then X_v, where $v = k(u,\mathscr{C})$, need not be a TVS. Most of what follows in this section will be found in H. S. Collins [5].

Definition 2. (*a*) Let E be a real vector space and u a topology on E. u is said to be *semi-linear* if, for each $x, y \in E$ and $\lambda \in R$ (real numbers), the mappings

$$(x,y) \to x-y,$$
$$(\lambda, x) \to \lambda x$$

are continuous in each variable separately. If those mappings are continuous in both variables together, then u is called *linear* or E_u, as usual, is a topological vector space.

(*b*) Let u be a linear (or semi-linear) topology on a vector space E. cu denotes the topology on E whose fundamental system of neighbourhoods of 0 consists of circled convex u-neighbourhoods of 0. Clearly, $u \supset cu$.

PROPOSITION 4. *If u is a semi-linear (or a linear) topology on a vector space E, then cu is a locally convex linear topology on E. In other words, E_{cu} is a locally convex topological vector space. Also cu is the finest locally convex topology coarser than u.*

Proof. The family $\{V\}$ forms a filter base for the neighbourhood system of 0 in the cu-topology where each V is the union of a collection of convex u-neighbourhoods U of 0. In view of ([3], Chapitre II, § 2, N° 2), $\{V\}$ defines a locally convex topology on E.

For the second part of the proposition, let v be a locally convex topology such that $u \supset v \supset cu$. That means for each convex circled V

in v there exists U in u such that $V \supset U$. Therefore the convex hull of V, being V, contains the convex hull of U. In other words, each V contains an element of the basis of cu. This proves that $v = cu$.

Proposition 5. *Let E_u be a TVS and $\mathscr{C}$ the collection of all u-compact subsets of E. Then the k-extension $k(u, \mathscr{C}) = v$ of u is a semi-linear topology.*

The proof is straightforward and therefore omitted. For details see ([5], Theorem 5).

Corollary 1. *Let E_u be a metrizable TVS and $\mathscr{C}$ the collection of all u-compact subsets of E. Then $u = k(u, \mathscr{C})$ is a linear topology.*

Proof. The fact that $k(u, \mathscr{C})$ is linear follows from its being equal to u, due to Proposition 1.

Since there do exist non-locally convex metrizable TVS (Chapter 1, Example 6), therefore cu which is the finest l.c. topology coarser than $u = k(u, \mathscr{C})$ in Corollary 1, is strictly coarser than $k(u, \mathscr{C})$. In other words, $k(u, \mathscr{C})$ need *not* be locally convex even though u is a linear topology.

Proposition 6. *Let E_u be a TVS and cu the finest locally convex topology coarser than u. Then $u \sim cu$* (Chapter 2, § 9, Def. 14), *and the equicontinuous sets in E' are the same with respect to u and cu.*

Proof. Since $u \supset cu$, $E'_u \supset E'_{cu}$. Let $x' \in E'_u$. There exists a u-neighbourhood U of 0 such that $\langle x, x' \rangle \leqslant 1$ for all $x \in U$. Since x' is linear, the same inequality is true if U is replaced by the convex hull of U. That means $x' \in E'_{cu}$. Hence $E'_u = E'_{cu}$.

For the second part, let H be a u-equicontinuous set in E'. Then $\{x : \langle x, x' \rangle \leqslant 1, x' \in H\}$, being convex, is a cu-neighbourhood of 0 and therefore H is a cu-equicontinuous set due to Proposition 14 (Chapter 2). However, it is obvious that each cu-equicontinuous set is u-equicontinuous.

3. The ew^*-topology and completeness

Definition 3. (*a*) Let E_u be an l.c. space and E' its dual. ew^* is defined to be the *finest topology* which coincides with $\sigma(E', E)$ on each equicontinuous set of E'.

In view of Proposition 14 (Chapter 2), the ew^*-topology can equivalently be described as follows: A subset W of E' is ew^*-open (ew^*-closed) if and only if $W \cap U^0$ is $\sigma(E', E)$-open ($\sigma(E', E)$-compact) in the relative topology of U^0 for each u-neighbourhood U of 0. In other words, ew^* is a k-extension of $\sigma(E', E)$ with respect to the class $\mathscr{C}$ which consists of polar sets U^0, where U runs over the u-neighbourhoods of 0 in E_u. Hence $ew^* \supset \sigma(E', E)$.

(*b*) Let cew^* denote the *finest locally convex topology* coarser than ew^*.

(1) By Proposition 5, ew^* is a semi-linear topology on E'. Clearly $ew^* \supset cew^*$.

PROPOSITION 7. *Let E' be the dual of an l.c. space E_u. Then*

$$ew^* \supset cew^* \supset p \supset k \supset c$$

(Chapter 2, § 9, Def. 16 (*a*)) *on E'.*

Proof. The proof is obvious because p as well as k is a locally convex topology and coincides with $\sigma(E', E)$ on each equicontinuous set of E' (Chapter 2, Prop. 11).

(2) In general each inclusion relation $ew^* \supset cew^* \supset p \supset k \supset c$ is strict, as will be shown later on for the first two inclusions. The remainder is well known.

PROPOSITION 8. *Let F be a dense subspace of an l.c. space E, and let F' and E' denote their duals endowed with the $\sigma(F', F)$- and $\sigma(E', E)$-topologies respectively. Then $\sigma(E', E)$ coincides with $\sigma(F', F)$ on each equicontinuous set of F'.*

Proof. Clearly $E' = F'$ algebraically. However, $\sigma(E', E) \supset \sigma(F', F)$ because each finite subset of F is a finite subset of E. In view of Proposition 14 (Chapter 2), it is sufficient to show that $\sigma(E,' E)$ coincides with $\sigma(F', F)$ on each U^0 for each neighbourhood U of 0 in F. Clearly $\overline{U}$ (closure taken in E) is a neighbourhood of 0 in E and therefore $U^0 = (\overline{U})^0$. Hence it follows that U^0 is compact in $\sigma(E', E)$ as well as in $\sigma(F', F)$. Therefore, by a general theorem in topology, it follows that $\sigma(E', E)$ coincides with $\sigma(F', F)$ on U^0.

PROPOSITION 9. *Let F be a dense subspace of an l.c. space E_u. Then the ew^*-topology on E' is the same with respect to E as with respect to F. In other words, F' and E', endowed with their respective ew^*-topologies, are homeomorphic.*

Proof. As in the previous proposition, for each neighbourhood U of 0 in F, $\overline{U}$ is a neighbourhood of 0 in E and $U^0 = (\overline{U})^0$. Hence for each set W,

$$W \cap U^0 = W \cap (\overline{U})^0.$$

Now since by the above proposition $\sigma(F', F)$ and $\sigma(E', E)$ coincide on U^0, therefore for W to be ew^*-open with respect to F is equivalent to being ew^*-open with respect to E.

THEOREM 1. *An l.c. space E_u is complete if and only if each linear functional on E' which is ew^*-continuous is $\sigma(E', E)$-continuous.*

Proof. (*a*) Suppose E_u is complete. Let F denote the set of all *ew**-continuous linear functionals on the dual E' of E. Clearly F is a vector subspace of the algebraic dual E'^* of E' and $E \subset F$. Let v denote the $\mathfrak{S}$-topology on F where $\mathfrak{S}$ consists of all convex circled $\sigma(E', E)$-closed equicontinuous sets of E'. Since for each $x \in F$ and M in $\mathfrak{S}$, $\langle x, M\rangle$, being continuous on M, is a bounded set of the reals, the $\mathfrak{S}$-topology is compatible with the vector space structure of F. Moreover, v induces u on E, because for each M in $\mathfrak{S}$, M^F (polar of M taken in F) is a v-neighbourhood of 0 in F by definition while M^E (polar of M taken in E) is a u-neighbourhood of 0 (Chapter 2, § 9, Prop. 14), and $M^F \cap E = M^E$. Since E_u is complete, E_u is a closed subspace of F_v. First we show that $F'_v = E'$. Clearly each M in $\mathfrak{S}$ is $\sigma(E', E)$-compact, and $\sigma(E', F) \supset \sigma(E', E)$ because $F \supset E$. But on the other hand, $\sigma(E', F)$ is the coarsest topology on E' in which each $x \in F$ is continuous on M. This shows that $\sigma(E', F) = \sigma(E', E)$ on each M in $\mathfrak{S}$. In other words, each M is $\sigma(E', F)$-compact. Hence by Theorem 6 (Chapter 2), $F'_v = E'_u$. Now, if $E \neq F$, by Theorem 2 (Chapter 2) there exists a v-closed hyperplane in F which contains E and separates each point $x \in F$, $x \notin E$. In other words, there exists a continuous functional $x' \neq 0$ in E' such that $\langle x, x'\rangle = 0$ for all $x \in E$. But this is not possible. Hence $E_u = F_v$.

(*b*) In order to prove the converse, let us suppose that the set of all *ew**-continuous linear functionals on E' coincides with E, i.e. $E'^{ew^*\prime} = E$. Let $\hat{E}_u$ denote the completion of E_u. Then $\hat{E}' = E'$ algebraically and E_u is dense in $\hat{E}$. Therefore by Proposition 9, $\hat{E}'^{ew^*} = E'^{ew^*}$ topologically and hence $\hat{E}'^{ew^*\prime v} = E'^{ew^*\prime v}$, where v is the $\mathfrak{S}$-topology described in (*a*). By part (*a*), $\hat{E}_u = \hat{E}'^{ew^*\prime v}$, and by assumption, $E'^{ew^*\prime v} = E_u$. Therefore $\hat{E}_u = E_u$. This completes the proof.

Theorem 1 can be paraphrased in such a fashion as to give a characterization of complete l.c. spaces analogous to that of B-complete spaces. Recall that non-zero linear functionals and hyperplanes are in one-to-one correspondence. Further, a linear functional f on E_u is continuous if and only if $f^{-1}(0)$ is closed. Moreover, a linear functional f on E' is *ew**-continuous if and only if $f^{-1}(0)$ is *ew**-closed. Thus we have:

THEOREM 2. *Let E_u be an l.c. space and E' its dual. The following statements are equivalent*:

(*a*) *E_u is complete.*

(*b*) *Each almost closed* (*in other words, ew*-closed*) *hyperplane of E' is $\sigma(E', E)$-closed.*

(*c*) *ew** $\sim$ *$\sigma(E', E)$ in the sense that they generate the same dual space.*

PROPOSITION 10. *Each B_r-complete (and therefore also B-complete) l.c. space E is complete. But the converse is not true.*

Proof. Let Q be an ew^*-closed hyperplane of the dual E' of E. Suppose Q is dense in E'. Then by Theorem 5 (Chapter 4, § 4), $Q = E'$, which is a contradiction. Hence Q must be $\sigma(E', E)$-closed. Therefore by Theorem 2, E is complete. The second part of the proposition follows from the example given in Proposition 1 (Chapter 4, § 1).

4. The ew^*-topology and pseudo-completeness

Definition 4. Let E_u be a TVS and E' its dual. E_u is said to be *pseudo-complete* if each ew^*-closed maximal proper subspace of E' is $\sigma(E', E)$-closed [5].

A pseudo-complete TVS is not necessarily locally convex. A connexion between pseudo-completeness and completeness and a detailed study of pseudo-completeness are due to Collins [5].

THEOREM 3. *Let E_u be a TVS. Then E_u is pseudo-complete if and only if E_{cu}* (Def. 2 (*b*)) *is complete.*

Proof. Suppose E_u is pseudo-complete. Let Q be an ew^*-closed hyperplane in E'_{cu}. Let x be the linear functional associated with Q. Then x is ew^*-continuous on $E'_u = E'_{cu}$ and hence $x^{-1}(0)$ is an ew^*-closed maximal linear subspace of E'. But then the assumption implies $x^{-1}(0)$ is $\sigma(E', E)$-closed. That means $x \in E'^{\sigma'} = E$, and hence Q is $\sigma(E', E)$-closed. In view of Theorem 2, this implies that E_{cu} is complete.

On the other hand, suppose E_{cu} is complete. Let Q be an ew^*-closed maximal linear subspace of E'_u. Let x be the linear functional associated with Q. Since $E'_u = E'_{cu}$ and the equicontinuous sets are the same with respect to u and cu, x is ew^*-continuous on E'_{cu}. But then the assumption implies x is $\sigma(E'_{cu}, E_{cu})$-continuous and hence $\sigma(E'_u, E_u)$-continuous. In other words, Q is $\sigma(E'_u, E_u)$-closed. This proves that E_u is pseudo-complete.

An immediate consequence of the above theorem is:

COROLLARY 2. *An l.c. space E_u is pseudo-complete if and only if it is complete.*

PROPOSITION 11. *Let v be any $\mathfrak{S}$-topology on the dual E' of an l.c. space E_u such that v coincides with $\sigma(E', E)$ on each equicontinuous set of E'. Then each M in $\mathfrak{S}$ is u-precompact. In other words, p is the finest $\mathfrak{S}$-topology which coincides with $\sigma(E', E)$ on each equicontinuous set of E'.*

Proof. To show $v \subset p$, let M be in $\mathfrak{S}$. Then M^0 is a v-neighbourhood of 0. Since v coincides with $\sigma(E', E)$ on each U^0, where U runs over

u-closed convex neighbourhoods of 0 in E_u, there exists a finite set F in E such that $M^0 \supset F^0 \cap U^0$. Hence $M \subset M^{00} \subset (F^0 \cap U^0)^0 =$ convex σ-closure of $(F^{00} \cup U) = F^{00}+U$, because the latter is closed and convex. But then it is easy to see that F^{00} is u-precompact. Therefore, there exists a finite subset G such that $M \subset F^{00}+U \subset G+2U$. In other words, M is u-precompact. Hence $v \subset p$.

Proposition 12. *Let E' be the dual of an l.c. space E_u. Let $cew^* = w$ denote the finest locally convex topology on E' which coincides with $\sigma(E', E)$ on each equicontinuous set of E'. Then $c(w) = p$* (Chapter 4, § 2).

Proof. Since c is an $\mathfrak{S}$-topology where $\mathfrak{S}$ consists of all convex circled u-compact subsets of E, $c \sim \sigma(E', E)$. Therefore a convex set is c-closed if and only if it is $\sigma(E', E)$-closed. Let $\mathfrak{S}'$ denote the collection of sets W^0, where W runs over convex circled c-closed w-neighbourhoods of 0 in E'. Clearly the $\mathfrak{S}'$-topology is $c(w)$, and $W^{00} =$ (σ-closure of W) $=$ (c-closure of W) $= W$. Therefore $c(w) \subset p$ by Proposition 11. On the other hand, if V is a p-neighbourhood of 0 in E', $V = P^0$ where P is a circled u-precompact set of E. Obviously V is a w-neighbourhood of 0 (because $w \supset p$), and being convex circled and σ-closed (therefore c-closed) it is a $c(w)$-neighbourhood of 0. This proves the proposition.

Proposition 13. *Let E_u be an l.c. space in which each u-closed u-precompact set is u-compact. Then $c = p$ on E'.*

Proof. In general $p \supset c$ always. For $p \subset c$, let V be a p-neighbourhood of 0. Then $V = P^0$ where P is u-precompact. Further, $V^0 = P^{00}$, and P^{00} being precompact and $\sigma(E, E')$-closed (therefore u-closed because $u \supset \sigma(E, E')$) implies P^{00} is u-compact due to hypothesis. Therefore V is a c-neighbourhood of 0. Hence $p = c$.

Corollary 3. *If E is a complete l.c. space, then $p = c$ on E'.*

Proof. It follows from Proposition 13 because in a complete l.c. space each closed precompact set is compact.

5. The *cew**-topology and completeness

Lemma 1. *Let E_u be an l.c. space and E' its dual. Let W be a convex ω-neighbourhood of 0 in E such that W^{00} is a u-neighbourhood, where ω denotes the finest locally convex topology on E. Then W is a u-neighbourhood of 0 if and only if each linear functional f which is bounded on W is in E'.*

Proof. The 'only if' part is trivial. For the 'if' part, let $p(x)$ be the semi-norm determined by W, i.e. $W \subset \{x : p(x) \leqslant 1\}$. By the Hahn–

Banach theorem, there exists a linear functional f such that $\langle x,f\rangle = p(x)$ for each $x \in E$. Hence $\sup_{x\in W}|\langle x,f\rangle| \leqslant 1$. Therefore f is bounded on W and hence, by hypothesis, $f \in E'$. But then $f \in W^0$, and the continuity of f implies $W^{00} \subset \{x: p(x) \leqslant 1\}$. Therefore $p(x)$ is a u-continuous semi-norm and hence $\{x: p(x) \leqslant 1\}$ is a u-neighbourhood of 0. But then $W \supset \{x: p(x) < 1\}$ implies W is a u-neighbourhood of 0.

The following theorem can be found in [5].

THEOREM 4. *An l.c. space E_u is complete if and only if the following conditions hold:*

(*a*) $cew^* = p$ *on* E' (*where* cew^* *is the finest locally convex topology coarser than* ew^*).

(*b*) *Each u-closed u-precompact set is u-compact.*

Proof. For the 'if' part, let $cew^* = w$. Since w is the finest locally convex topology coarser than ew^*, $w \sim ew^*$ by Proposition 6. Further, condition (*b*) implies $p = c$ by Proposition 13 and $c \sim \sigma(E', E)$ due to Theorem 6 (Chapter 2). Hence $ew^* \sim \sigma(E', E)$. But this proves that E_u is complete due to Theorem 2.

The 'only if' part: If E_u is complete then part (*b*) of the theorem is clearly true. Further, $w \supset p$ in general. For $w \subset p$, let W be a convex circled w-neighbourhood of 0 in E'. Clearly W is an ω'-neighbourhood of 0 because ω' is the finest locally convex topology in E'. Moreover W^{00}, being a $c(w)$-neighbourhood of 0, is a p-neighbourhood of 0 by Proposition 12. Now if f is a linear functional which is bounded on W, then $f \in E'^{w'}$. But $w \sim ew^*$ implies $f \in E'^{ew^{*\prime}}$. Since E_u is complete by hypothesis, $ew^* \sim \sigma(E', E) \sim c = p$ (the first equivalence is due to Theorem 2, the middle one is due to Theorem 6 (Chapter 2), and the last equality is due to Corollary 3). Therefore $f \in E'^{p'}$. But then Lemma 1 shows that W is a p-neighbourhood of 0. This proves part (*a*), and thus the theorem is completed.

COROLLARY 4. *Let E' be the dual of a complete l.c. space E_u. Then, on E',*

$$ew^* \sim cew^* = p = k = c \sim \sigma(E', E).$$

Proof. All relations except $p = k$ have been accounted for in Theorem 4. $p = k$ follows from the fact that $p \supset k \supset c$ in general and $p = c$ due to Corollary 3.

PROPOSITION 14. *Let* $E = \prod_\alpha R_\alpha$ $(\alpha \in A)$ *be an arbitrary cartesian product with the product topology, where R_α is a copy of the reals. Then E is B-complete.*

Proof. Clearly E is complete and also pseudo-complete (Corollary 2). Further $E' = \sum_{\alpha \in A} R_\alpha$. First we show that every subspace of E' is *ew**-closed. Let Q be any subspace of E'. Since each convex $\sigma(E', E)$-closed equicontinuous set H of E' is contained in the convex closure M of a finite set F (say), therefore

$$Q \cap H = Q \cap M \cap H = Q \cap L(M) \cap M \cap H,$$

where $L(M)$ is the vector subspace generated by F. Since $Q \cap L(M)$ is a finite-dimensional vector space (because $L(M)$ is finite-dimensional), it is closed due to Corollary 3 (Chapter 3, § 1) and therefore $Q \cap H$ is $\sigma(E', E)$-closed on H because M is $\sigma(E', E)$-closed.

Now any arbitrary vector subspace Q of E' can be written as the intersection of maximal proper vector subspaces each containing Q. Since each maximal vector subspace, being *ew**-closed (because of the previous paragraph), is $\sigma(E', E)$-closed because E_u is pseudo-complete, therefore Q, being the intersection of σ-closed subspaces, is σ-closed. This proves that E is B-complete.

The following proposition is a well-known result due to S. Kaplan [17]. But the proof given here is due to V. Pták [29].

PROPOSITION 15. *Let E_ω be an l.c. space with the finest locally convex topology. Then E_ω is complete.*

Proof. Observe that $\omega = \tau(E, E^*)$, where E^* is the algebraic dual of E. Let e_α $(\alpha \in A)$ be a Hamel basis of E and let e'_β $(\beta \in A)$ denote a subset of E^* such that $\langle e_\alpha, e'_\beta \rangle = 0$ or 1 according as $\alpha \neq \beta$ or $\alpha = \beta$.

Let $\hat{x}_0 \in \hat{E}_\omega$. We show that $\hat{x}_0 \in E$. Let B denote a subset of A such that $\langle \hat{x}_0, e'_\beta \rangle \neq 0$ $(\beta \in B)$. Suppose $B \neq \emptyset$. Define

$$p(x) = \sup_{\beta \in B} \frac{\langle x, e'_\beta \rangle}{\langle \hat{x}_0, e'_\beta \rangle}.$$

p is a semi-norm on E and therefore also on $\hat{E}$. Since E is dense in $\hat{E}$, there exists $x_0 \in E$ such that $p(x_0 - \hat{x}_0) \leqslant \frac{1}{2}$. Therefore

$$p(x_0) = p(x_0 - \hat{x}_0 + \hat{x}_0) \geqslant p(\hat{x}_0) - p(x_0 - \hat{x}_0) \geqslant \tfrac{1}{2}.$$

Hence $\langle x_0, e'_\beta \rangle \neq 0$ for every $\beta \in B$. But this means that B must be finite. Thus for $x_0 \in E$ we have

$$\langle \hat{x}_0, e'_\alpha \rangle = \langle x_0, e'_\alpha \rangle \quad \text{for every } \alpha \in A.$$

From this it follows that

$$\langle \hat{x}_0, x' \rangle = \langle x_0, x' \rangle$$

where x' is an element in the subspace Q generated by e'_α $(\alpha \in A)$ in E^*. Let y'_0 be an arbitrary element of E^*. The set

$$\{y' \in E^*: |\langle e'_\alpha, y'\rangle| \leqslant |\langle e'_\alpha, y'_0\rangle|, \alpha \in A\}$$

is a U^0, where U is an ω-neighbourhood of 0. Since Q is dense in U^0, and $\hat{x}_0$ is continuous on U^0 and is equal to x_0 on a dense subset of U^0, we have $\hat{x}_0 = x_0$ on U^0. Therefore $\langle \hat{x}_0, y'_0\rangle = \langle x_0, y'_0\rangle$. But this proves that $\hat{x}_0 = x_0$, because y'_0 was arbitrary. Hence $\hat{x}_0 \in E$.

The following two examples reveal that $ew^* \supset cew^* \supset p$ are strict inclusions in general. They are due to Collins [5].

Example 1. The fact that the inclusion $ew^* \supset cew^*$ is strict follows from the example of E_ω considered in Proposition 1 (Chapter 4). For, E_ω being complete (Prop. 15), $\sigma(E', E) \sim c = p = cew^*$ on E'_ω by Corollary 4. However, $ew^* \neq cew^*$ because otherwise each ew^*-closed subspace, being convex, would be $\sigma(E'_\omega, E_\omega)$-closed. In other words, E_ω would be B-complete which is not true (Chapter 4, § 1, Prop. 1).

Example 2. To show that $cew^* \supset p$ is a strict inclusion, we consider the dual E'^σ of an infinite-dimensional Banach space E_u. Clearly, $p = u$ on E_u because each $\sigma(E', E)$-precompact set, being $\sigma(E', E)$-bounded, is equicontinuous (Chapter 2, § 9, Theorem 7), since every Banach space is a t-space. Moreover, each $\sigma(E', E)$-closed and $\sigma(E', E)$-precompact set is $\sigma(E', E)$-compact in E'. Now if $p = cew^*$, then it would follow that E'^σ is complete by Theorem 4. But this is clearly not true. Hence $cew^* \supset p$ is a strict inclusion.

6. The ew^*-topology and hypercompleteness

The significance of the ew^*-topology on the dual E' of an l.c. space E_u has already been shown in the previous sections of this chapter. Another class of topological vector spaces which are called 'hypercomplete'—a notion due to J. L. Kelley [19]—can also be described with the help of the ew^*-topology.

Let $\mathscr{G}$ denote the family of all subsets of an l.c. space E_u. Then one can define a uniform structure on $\mathscr{G}$ as follows: For each neighbourhood U of 0, let $W_U = \{(A, B): A \subset B+U \text{ and } B \subset A+U, A, B \text{ in } \mathscr{G}\}$ denote the collection of sets. It is easily seen that the totality of sets W_U forms a filterbase of a uniformity on $\mathscr{G}$. $\mathscr{G}$ is said to be complete if it is complete under this uniform structure.

Definition 5. (*a*) Let E_u be an l.c. space. Let $\mathscr{C}$ denote the class of all convex circled subsets of E. If $\mathscr{C}$ is complete in the above sense, then E_u is said to be *hypercomplete*.

(*b*) Let $\mathscr{K} = \{K\}$ be a filterbase in an l.c. space E_u, where each K is convex and circled. Then $\mathscr{K}$ is said to be *fundamental* if for each neighbourhood U of 0 in E, there exists K' in $\mathscr{K}$ such that $K' \subset K+U$ for each K in $\mathscr{K}$.

(*c*) A fundamental filterbase $\mathscr{K} = \{K\}$ is said to be *convergent* if, for each neighbourhood U of 0 in E, there exists K' in $\mathscr{K}$ such that $K' \subset C+U$, where $C = \bigcap_{K \in \mathscr{K}} \bar{K}$.

THEOREM 5. *Let E_u be an l.c. space and E' its dual. Then the following statements are equivalent*:

(*a*) *E_u is hypercomplete.*

(*b*) *Each fundamental filter base $\mathscr{K} = \{K\}$ in E, where each K is convex and circled, converges in E.*

(*c*) *Each ew*-closed convex circled subset of E' is $\sigma(E', E)$-closed.*

For the proof see Kelley [19].

COROLLARY 5. *Every hypercomplete l.c. space is B-complete and therefore complete.*

Proof. Since every subspace is convex and circled, the corollary follows from Theorem 5 (*c*) and Theorem 1 (Chapter 4).

An example of a B-complete l.c. space which is non-hypercomplete is *not* known. In the next chapter we shall study an important class of l.c. spaces which are hypercomplete.

PROPOSITION 16. (*a*) *A closed subspace M of a hypercomplete space E is hypercomplete.*

(*b*) *If M is a closed subspace of a hypercomplete l.c. space E, then E/M is hypercomplete.*

Proof. (*a*) follows from Proposition 4 (Chapter 4) by replacing a subspace Q by a convex circled set of the dual M' of M;

(*b*) follows from the fact that $(E/M)' = M^0 \subset E'$ and each convex circled *ew**-closed subset C of M^0 is *ew**-closed in E', and therefore it is $\sigma(E', E)$-closed because E is hypercomplete. Hence C is *a fortiori* $\sigma(M', M)$-closed. This completes the proof.

(1) (*a*) Since every hypercomplete space is B-complete (Corollary 5), an arbitrary direct product and sum of hypercomplete spaces is not hypercomplete, due to the counter examples given after Corollary 2, § 1 (Chapter 4). (*b*) A counter example due to Grothendieck [11] shows that even a countable direct sum of Fréchet spaces is not B-complete and therefore not hypercomplete. (*c*) The question whether a finite

sum of hypercomplete spaces is hypercomplete, remains open. (*d*) It is also not known if for each l.c. space E there exists a hypercompletion $\hat{E}^h$ or a B-completion $\hat{E}^b$. (*e*) In the next chapter we shall discuss a class of l.c. spaces the completion of each of which is hypercomplete and therefore B-complete.

Exercises

(1) Let E_u be an l.c. space and $\hat{E}_u$ its completion. Show that:

(*a*) $\hat{E}_u$ is hypercomplete if and only if each ew^*-closed convex circled subset C of E' is the intersection of half-spaces containing C such that each hyperplane associated with each half-space is ew^*-closed. (See [36].)

(*b*) $\hat{E}_u$ is B-complete if and only if each ew^*-closed subspace of E' is the intersection of ew^*-closed hyperplanes containing it. (See [32].)

(2) Let $\mathscr{M}$ denote the class of all metrizable l.c. or normed spaces. Show that the completion of each $B(\mathscr{M})$-space (Chapter 7, § 1, Def. 1) is also a $B(\mathscr{M})$-space. (See [36].)

(3) Let E_u be a hypercomplete l.c. space and F_v a t-space.

(*a*) A linear continuous mapping of E_u onto F_v is open.

(*b*) A linear mapping of F_v into E_u with the closed graph is continuous.

(4) An l.c. space E_u is said to be *minimal* if there exists no locally convex Hausdorff topology (on E) strictly coarser than u. Let E_u be a minimal space and F_v any l.c. space.

(*a*) A linear continuous mapping of E onto F is open.

(*b*) A linear mapping of F into E with the closed graph is continuous.

(*c*) Every minimal l.c. space is B_r-complete.

6

THE THEORY OF S-SPACES

1. Definition and a characterization of S-spaces

THE theory of S-spaces is due to T. Husain [14]. Some of the important properties of S-spaces are the following: (*a*) they are not necessarily metrizable, although every metrizable l.c. space is an S-space; (*b*) the Krein–Šmulian theorem which is true for Fréchet spaces can also be proved for complete S-spaces; (*c*) the completion of an S-space is B-complete; (*d*) every subspace of an S-space E is an S-space provided E satisfies a closure property (see the main text); (*e*) the dual E'^c of a complete S-space E is B_r-complete, provided E satisfies the closure property.

All of the results given in this chapter can be found in Husain's paper [14].

Definition 1. Let ew^* and p (Chapter 5, § 3, Def. 3 (*a*) and Chapter 2, Def. 16) be the usual topologies on the dual E' of an l.c. space E_u. E_u is said to be an *S-space* if $ew^* = p$.

Remark. In general, ew^* is not a linear topology, and $ew^* \supset p$. But for S-spaces ew^* is a locally convex topology because p is so.

THEOREM 1. *An l.c. space E_u is an S-space if and only if for each ew^*-open neighbourhood W' of 0 in E', there exists a u-precompact set P in E_u such that $P^0 \subset W'$.*

Proof. If E_u is an S-space then $ew^* = p$ on E' and therefore the condition of the theorem is obviously satisfied. On the other hand if, for each ew^*-open neighbourhood W' of 0, there exists a u-precompact set P in E such that $P^0 \subset W'$, then $ew^* \subset p$. But in general $ew^* \supset p$ (Chapter 5, Prop. 7), and hence, combining the two inequalities, we have $ew^* = p$. This proves that E is an S-space.

The first part of the following theorem is well known ([4], Chapitre IV, § 2, N° 6, Prop. 7) but not in the form given here.

THEOREM 2. *Every metrizable l.c. space E_u is an S-space, but the converse is not true.*

Proof. Let U_n ($n \geqslant 1$) denote a monotonically decreasing sequence of convex circled and closed neighbourhoods of 0 in E_u, and let W' be an arbitrary ew^*-open neighbourhood of 0 in E'. We wish to show first of all that there exist a finite subset $A_1 \subset E$ and, for $n > 1$, a finite subset $B_{n-1} \subset U_{n-1}$, $A_n = A_{n-1} \cup B_{n-1}$ such that $A_n^0 \cap U_n^0 \subset W'$ for all $n \geqslant 1$.

Since U_1^0 is equicontinuous and ew^* coincides with $\sigma(E', E)$ on U_1^0, there exists a finite subset $A_1 \subset E$ such that $A_1^0 \cap U_1^0 \subset W'$.

Now suppose, by induction, that there exists a finite subset $B_{n-1} \subset U_{n-1}$, $A_n = A_{n-1} \cup B_{n-1}$ such that $A_n^0 \cap U_n^0 \subset W'$, and assume that there exists no finite subset $B_n \subset U_n$ such that $A_{n+1} = A_n \cup B_n$ implies $A_{n+1}^0 \cap U_{n+1}^0 \subset W'$. In other words, for each finite subset $B \subset U_n$, $(A_n \cup B)^0 \cap K_n' = A_n^0 \cap B^0 \cap K_n' \neq \emptyset$, where $K_n' = U_{n+1}^0 \cap (E' \setminus W')$. Since $E' \setminus W'$ and U_{n+1}^0 are ew^*-closed and since ew^* coincides with $\sigma(E', E)$ on U_{n+1}^0, therefore K_n', being a $\sigma(E', E)$-closed subset of a $\sigma(E', E)$-compact subset U_{n+1}^0, is $\sigma(E', E)$-compact. Furthermore, the collection of sets $A_n^0 \cap B^0 \cap K_n'$ forms a filterbase when B runs over the family $\mathfrak{S}$ of finite subsets of U_n, because $A_n^0 \cap B^0 \cap K_n' \neq \emptyset$ for each $B \in \mathfrak{S}$ by assumption. Clearly each $A_n^0 \cap B^0 \cap K_n'$ is $\sigma(E', E)$-closed, and therefore compactness of K_n' implies there exists $x' \in A_n^0 \cap B^0 \cap K_n'$ for each finite subset B of U_n. But $U_n = \bigcup B$, B in $\mathfrak{S}$, and therefore $x' \in A_n^0 \cap U_n^0 \cap K_n'$. On the other hand, by the assumption due to induction, $A_n^0 \cap U_n^0 \subset W'$ implies $A_n^0 \cap U_n^0 \cap K_n' \subset W' \cap K_n' = \emptyset$, because $K_n' \subset E' \setminus W'$. Hence $x' \in \emptyset$ is an absurdity and therefore for each $n > 1$ there exists a finite subset $B_{n-1} \subset U_{n-1}$ such that if

$$A_n = A_{n-1} \cup B_{n-1}$$

then $A_n^0 \cap U_n^0 \subset W'$ for all $n \geqslant 1$.

Let $P = \bigcup_{n \geqslant 1} A_n$. We show that P is u-precompact. For this let U be an arbitrary u-neighbourhood of 0. Since U_n $(n \geqslant 1)$ is a monotonically decreasing sequence of u-neighbourhoods of 0, there exists a positive integer n_0 such that for $n \geqslant n_0$, $B_n \subset U_n \subset U_{n_0} \subset U$. Let $B = \bigcup_{n < n_0} (B_n \cup A_1)$. B is a finite subset of E, and

$$P = \bigcup_{n \geqslant 0} A_{n+1} = \bigcup_{n \geqslant 0} (B_n \cup A_n)$$

$$\subset \bigcup_{n \geqslant n_0} (B_n \cup B) \subset U_{n_0} \cup B \subset U + B.$$

This proves that P is u-precompact.

Further, $P \supset A_n$ for each n implies $P^0 \subset A_n^0$ and therefore

$$P^0 \cap U_n^0 \subset A_n^0 \cap U_n^0 \subset W'.$$

But
$$E' = \bigcup_{n \geqslant 1} U_n^0 = (\bigcap_{n \geqslant 1} U_n)^0$$

and hence

$$P^0 = P^0 \cap E' = P^0 \cap \left(\bigcup_{n\geqslant 1} U_n^0\right) = \bigcup_{n\geqslant 1} (P^0 \cap U_n^0) \subset W'.$$

By Theorem 1, this proves that E_u is an S-space.

For the converse, consider $E = R^{(N)}$ (Chapter 2, § 1, Example 4) with the finest locally convex topology ω. It is easy to see that E_ω is the strict inductive limit of a strictly increasing sequence of finite-dimensional Euclidean spaces R^n $(n \geqslant 1)$ which are Fréchet spaces. By Proposition 15 (Chapter 5), E_ω is complete. E_ω is *not* metrizable because otherwise E_ω would be a Fréchet space and hence of the second category by Baire's Theorem (Chapter 1); but E_ω is a countable union of non-dense subsets (because the identity mapping: $R^n \to R^{n+1}$ is 'into' for each n by Corollary 5 (Chapter 3, § 2)) and therefore of the first category. This is a contradiction and therefore E_ω is not metrizable.

To show that E_ω is an S-space, consider $E'_\omega = R^N$ (the countable product of the reals). Since E_ω is complete, the topologies p and c on E' are equal due to Corollary 3 (Chapter 5, § 4). Moreover, $c = \sigma(E'_\omega, E_\omega)$ on E' because every *convex* compact set in E_ω is finite-dimensional (Chapter 2, § 6, Prop. 8). But $\sigma(E'_\omega, E_\omega)$ is simply the product topology in which R^N is metrizable and therefore by Corollary 1 (Chapter 5, § 2), $ew^* = \sigma(E'_\omega, E_\omega)$. Combining the latter with the above equation we get $ew^* = p$. This proves that E_ω is an S-space.

2. S-spaces and B-completeness

Theorem 2 of § 1 shows that an S-space is not necessarily complete. Nor is every complete space an S-space (Chapter 5, § 5, Example 1 following Proposition 15). The following theorem shows that with some additional conditions on an S-space one has B-completeness.

THEOREM 3. *An S-space E_u is B-complete provided each u-closed and u-precompact set in E_u is u-compact.*

Proof. By the definition of an S-space, $ew^* = cew^* = p$ on E'. Further, the hypothesis—that every u-closed and u-precompact set is compact—coupled with '$cew^* = p$' implies E_u is complete by Theorem 4 (Chapter 5, § 5), and hence $ew^* = cew^* = p = c \sim \sigma(E', E)$ on E' by Corollary 4 (Chapter 5, § 5).

Now let Q be any ew^*-closed subspace of E'. Then in view of the previous paragraph Q, being convex, is $\sigma(E', E)$-closed due to Proposition 15 (Chapter 2, § 9). Hence E_u is B-complete due to Theorem 1 (Chapter 4, § 1).

Since in a complete space each closed precompact set is compact, therefore the following corollary is immediate from Theorem 3.

COROLLARY 1. *A complete S-space is B-complete.*

PROPOSITION 1. *An S-space E_u is complete if and only if every u-closed and u-precompact set is u-compact.*

Proof. If E_u is an S-space which satisfies the condition in the proposition, then E_u is B-complete due to Theorem 3 and therefore complete by Proposition 10 (Chapter 5, § 3). The other part is trivial.

PROPOSITION 2. *A complete l.c. space E_u is an S-space if and only if ew^* is a locally convex topology on E'.*

Proof. The fact that ew^* is locally convex for S-spaces has been pointed out in the remark following Definition 1 (§ 1). On the other hand, if ew^* is locally convex, then $ew^* = cew^*$. But E being complete, $cew^* = p$ on E' by Theorem 4 (Chapter 5, § 5). Therefore $ew^* = p$ and hence E is an S-space.

COROLLARY 2. *An LF-space E is B-complete if ew^* on E' is locally convex.*

Proof. It follows from Proposition 2 because each LF-space is complete.

THEOREM 4. *The completion $\hat{E}_u$ of an S-space E_u is an S-space.*

Proof. We can identify the duals E'_u and $\hat{E}'_u$ of E_u and $\hat{E}_u$ respectively under the mapping: $f \to \hat{f}$, where $f \in E'$ and $\hat{f}$ is the unique extension of f. By Proposition 9 (Chapter 5, § 3), E'^{ew^*} and $\hat{E}'^{\widehat{ew}^*}$ are homeomorphic (where ew^* and $\widehat{ew}^*$ are usual topologies on E' and $\hat{E}'$ respectively) because E is dense in $\hat{E}$. Let p and $\hat{p}$ denote the $\mathfrak{S}$-topologies on E' and $\hat{E}'$ where $\mathfrak{S}$ consists of precompact sets of E and $\hat{E}$ respectively. Since the identity mapping: $E_u \to \hat{E}_u$ is continuous, each u-precompact set in E is u-precompact in $\hat{E}_u$. Therefore $\hat{p} \supset p$. But in general, $\widehat{ew}^* \supset \hat{p} \supset p$, and by hypothesis, $\widehat{ew}^* = ew^* = p$. Therefore $\widehat{ew}^* = ew^* = p = \hat{p}$ shows that $\hat{E}_u$ is an S-space.

COROLLARY 3. *The completion $\hat{E}_u$ of an S-space E_u can be obtained by completing the precompact sets of E_u only.*

Proof. This is immediate from the above theorem.

COROLLARY 4. *If E_u is an S-space then $\hat{E}_u$ is B-complete.*

Proof. The fact that $\hat{E}_u$ is B-complete follows from Theorem 4 coupled with Corollary 1.

THEOREM 5. *Let E_u be an S-space and F_v a t-space. Then:*

(*a*) *A linear and continuous mapping of E_u onto F_v is open provided the filter condition holds* (Chapter 4, § 2, Def. 4).

(*b*) *A linear mapping g of F_v into E_u, the graph of which is closed in $F \times E$, is continuous provided the inverse filter condition holds* (Chapter 4, § 3, Def. 6).

Proof. The proofs of (*a*) and (*b*) are exactly like the proofs of Theorems 3 and 4 of Chapter 4, because $\hat{E}_u$ is B-complete by Corollary 4 and therefore an application of Theorem 8 (Chapter 4, § 5) enables us to reproduce the details of Theorems 3 and 4 of Chapter 4, § 3.

It is worth noting that Corollary 4 describes a situation which is quite familiar for metrizable l.c. spaces—namely the completion of a metrizable l.c. space is a Fréchet space. However, it is not known if every B-complete space is the completion of some S-space. Nor is it known if there exists a B-complete l.c. space which is not a complete S-space.

Due to Proposition 1 (Chapter 4, § 1), a complete l.c. space is not necessarily an S-space because E_ω in the example considered in that proposition is complete but not B-complete and hence not an S-space. Actually, in view of the remark before Corollary 1 (Chapter 4, § 1), a complete bornological t-space need not be an S-space.

3. The Krein-Šmulian theorem

One of the main reasons for studying S-spaces is to generalize the so-called Krein–Šmulian theorem which is known to be true for Fréchet spaces. The following theorem is the most general form, so far, of the Krein–Šmulian theorem.

THEOREM 6. (*Husain* [14].) *Let E_u be an S-space in which each u-closed and u-precompact set is u-compact. Then a convex set M' in E' is $\sigma(E', E)$-closed if and only if it is ew^*-closed.*

Proof. The proof of this theorem is exactly like that of Theorem 3. By hypothesis, it follows that the following relations hold between the topologies on E', namely,

$$ew^* = cew^* = p = c \sim \sigma(E', E).$$

Hence convex closed sets in the topologies ew^* and $\sigma(E', E)$ are the same due to Proposition 15 (Chapter 2, § 9).

COROLLARY 5. (*Krein–Šmulian.*) *Let E_u be a Fréchet* (*or Banach*) *space. Then a convex set M' of E' is $\sigma(E', E)$-closed if and only if, for each u-neighbourhood U of 0 in E, $M' \cap U^0$ is $\sigma(E', E)$-closed.*

Proof. The corollary follows from Theorem 6 by observing that to say $M' \cap U^0$ is $\sigma(E', E)$-closed for each u-neighbourhood U of 0 in E is equivalent to saying that M' is ew^*-closed, and that each Fréchet space, being metrizable, is an S-space (Theorem 2), and that its being complete implies each u-closed u-precompact set in E_u is u-compact.

COROLLARY 6. *Every complete S-space E is hypercomplete.*

Proof. Let M' be any convex circled and ew^*-closed subset of E'. Then M' is $\sigma(E', E)$-closed by Theorem 6. Hence E_u is hypercomplete due to Theorem 5 (Chapter 5, § 6).

It is not known if the converse is true.

4. Subspaces and quotient spaces of an S-space

PROPOSITION 3. *Let E_u be an S-space in which each u-closed and u-precompact set is u-compact. Then every closed subspace Q of E is a complete S-space and therefore* a fortiori *B-complete.*

Proof. By Theorem 3, E_u is B-complete and therefore complete. Since a closed subspace of a complete l.c. space is complete, in order to prove the proposition it is sufficient to show that $ew_1^* = c_1$ on $Q' = E'/Q^0$, where ew_1^* and c_1 are the usual topologies on Q'. It is easy to see that c_1 and the quotient topology c^q of c both on Q' are equal to each other ([4], Chapitre IV, § 3, N° 4, Ex. 12 (c)). But $ew^* = c$ on E' (because E is a complete S-space) implies $ew^{*q} = c^q$. Therefore we have $ew^{*q} = c^q = p_1 = c_1$ on Q'.

Furthermore, ew_1^* being the *finest* topology which coincides with $\sigma(Q', Q)$ on each equicontinuous set of Q', $ew_1^* \supset ew^{*q} = c_1$ (observe that $ew^{*q} = c_1$ shows that ew^{*q} coincides with $\sigma(Q', Q)$ on each equicontinuous set of Q' (Chapter 2, § 8, Prop. 8)). In order to show the reverse inclusion, namely $ew_1^* \subset ew^{*q}$, let W' be an ew_1^*-open neighbourhood of 0 in Q'. For each u-neighbourhood U of 0 in E_u, consider

$$\begin{aligned}\varphi^{-1}(W') \cap U^0 &= \varphi^{-1}(W') \cap \varphi^{-1}(\varphi(U^0)) \cap U^0 \\ &= \varphi^{-1}(W' \cap \varphi(U^0)) \cap U^0\end{aligned}$$

because $U^0 \subset \varphi^{-1}(\varphi(U^0))$ and φ is the canonical mapping: $E' \to E'/Q^0$. Since φ is weakly continuous, $\varphi(U^0)$ is a convex compact set of Q', and therefore $W' \cap \varphi(U^0)$ is relatively $\sigma(Q', Q)$-open on $\varphi(U^0)$. This shows $\varphi^{-1}(W' \cap \varphi(U^0)) \cap U^0$ is relatively $\sigma(E', E)$-open on U^0 for each U. This proves that $\varphi^{-1}(W')$ is ew^*-open. Hence $ew_1^* = ew^{*q} = p_1 = c_1$ completes the proof.

Definition 2. A TVS E is said to satisfy CP (*Closure-Property*) if for any dense subspace Q of E, each precompact set of E is contained in the closure of a precompact set of Q.

In general an S-space need not satisfy CP. Every metrizable l.c. space, of course, does so.

PROPOSITION 4. *Let E_u be an S-space satisfying CP. Then every dense subspace Q of E is an S-space.*

Proof. Density of Q implies $Q' = E'$. By Proposition 9 (Chapter 5, § 3), $ew^* = ew_1^*$, where ew^* and ew_1^* are the usual topologies defined on E' with respect to E and Q respectively. By hypothesis, for each precompact set P in E there exists a precompact set P_1 in Q such that $\overline{P}_1 \supset P$. This proves that $p = p_1$, where p and p_1 are usual $\mathfrak{S}$-topologies on E' where $\mathfrak{S}$ consists of precompact sets of E and Q respectively. Therefore, $ew_1^* = ew^* \supset p = p_1$. Since E is an S-space by hypothesis, $ew_1^* = p_1$ implies Q is also an S-space.

Combining Propositions 3 and 4, we have an interesting property of S-spaces which is also shared by metrizable l.c. spaces.

THEOREM 7. *Let E_u be an S-space satisfying CP and in which each u-closed u-precompact set is u-compact. Then every subspace Q of E is an S-space.*

Proof. According to Proposition 3, $\overline{Q}$ is an S-space in which Q is dense, and it satisfies CP. Therefore, by Proposition 4, Q is an S-space.

THEOREM 8. *Let f be a linear continuous and almost open mapping of an l.c. space E onto another l.c. space F. If E is an S-space then F is also an S-space.*

Proof. Let f' denote the transpose mapping: $F' \to E'$. By hypothesis, f' is a homeomorphism (into). Let F' be identified with its image $f'(F')$. Let W' be an ew^*-open neighbourhood of 0 in F', i.e. $W' \cap V^0$ is $\sigma(F', F)$-open for each neighbourhood V of 0 in F. Consider

$$\begin{aligned} W' \cap U^0 &= W' \cap f'f'^{-1}(U^0) = W' \cap f'(f(U))^0 \\ &= W' \cap f'(\overline{f(U)})^0 = W' \cap f'(V^0) \\ &= W' \cap V^0. \end{aligned}$$

Since f is almost open, $\overline{f(U)} = V$ is a neighbourhood of 0 in F and therefore $W' \cap V^0$ being $\sigma(F', F)$-open implies $W' \cap U^0$ is $\sigma(E', E)$-open on U^0, for each U. In other words, W' is ew^*-open in E'. Since E is an S-space, there exists a u-precompact set P in E such that $P^0 \subset W'$ by Theorem 1. But then

$$(f(P))^0 = f'^{-1}(P^0) \subset f'^{-1}(W') = W'$$

implies F is an S-space, due to Theorem 1 because f being continuous and P being u-precompact implies $f(P)$ is precompact in F.

COROLLARY 7. *Let E and F be two l.c. spaces and f a linear continuous and open mapping of E onto F. If E is an S-space then F is also an S-space.*

Proof. The corollary is immediate from Theorem 8 because every open mapping is almost open.

COROLLARY 8. *Let E be an S-space and M a closed subspace of E. Then E/M is an S-space.*

Proof. This is a particular case of Corollary 7 because the canonical mapping: $E \to E/M$ is linear continuous open and onto.

COROLLARY 9. *Let E_u be an l.c. space and v another locally convex topology on E such that $u \supset v$ and $v(u) = v$. Then E_v is an S-space if E_u is an S-space.*

Proof. This corollary follows from Theorem 8, because $u \supset v$ and $v(u) = v$ are, respectively, equivalent to the continuity and almost openness of the identity mapping: $E_u \to E_v$.

5. The dual of an S-space

We know (Chapter 4, § 1, Prop. 7) that the dual of a Fréchet space, endowed with the c-topology, is B-complete. Before showing that the dual of a complete S-space satisfying CP is also B-complete, we first exhibit that the dual of a B-complete l.c. space under the same topology is not necessarily B-complete. For example, consider $E = \prod_{\alpha \in A} R_\alpha$ (an arbitrary product of the reals) endowed with the product topology. According to Proposition 14 (Chapter 5, § 5), E is B-complete. Further, each closed bounded set in E is compact. Therefore the topologies c and β on E' are equal. As β is the direct sum topology of finite-dimensional spaces, it is the finest locally convex topology, and $E'^{\beta} = E'^{c} = \sum_{\alpha \in A} R_\alpha$. However, in view of Proposition 1 (Chapter 4, § 1), it is easy to see that E'^{c} is *not* B-complete.

THEOREM 9. *Let E_u be a complete S-space satisfying CP* (§ 4, Def. 2). *Then E'^{v} is B_r-complete for any locally convex topology v such that $c \subset v \subset \tau(E', E)$.*

Proof. In view of Proposition 6 (Chapter 4, § 1), it is sufficient to show that E'^{c} is B_r-complete.

Let Q be a dense subspace of E such that $Q \cap C^{00}$ is $\sigma(E, E')$-closed for each u-compact convex set C of E. The density of Q implies $Q' = E'$ and that the ew^*-topology is the same with respect to E and Q due to Proposition 9 (Chapter 5, § 3). In view of Proposition 4 (§ 4), by hypothesis it follows that Q is an S-space, i.e. $ew^* = p_1$, where p_1 is the $\mathfrak{S}$-topology on E' where $\mathfrak{S}$ consists of precompact sets of Q.

We wish to show that every closed precompact set of Q is compact. For this, let c and c_1 be the usual topologies with respect to E and Q on E'. Since every compact set of Q is compact in E, $c_1 \subset c$; and $E'^{c'} = E$ and $E'^{c_1'} = Q$. Since Q is almost closed in E, the identity mapping: $E'^{c} \to E'^{c_1}$ is almost open. Or in other words, $c_1(c) = c_1$ (Chapter 4, Props. 9 and 14). But in general, by Proposition 12 (Chapter 5, § 4), $c_1(w_1) = p_1$, where $w_1 = cew_1^*$ is the finest locally convex topology which coincides with $\sigma(Q', Q)$ on each equicontinuous set of Q'. But Q is an S-space, and therefore $c_1(ew^*) = c_1(w_1) = p_1$. Moreover, E is a complete S-space, so $ew^* = p = c$ implies

$$c_1 = c_1(c) = c_1(ew^*) = c_1(w_1) = p_1.$$

This proves the assertion that every closed precompact set of Q is compact. Hence, by Proposition 1, Q is complete and therefore closed in E. Since Q is already dense in E, $Q = E$. In view of Theorem 5 (Chapter 4, § 4), this completes the proof.

Theorem 10. *The strong dual* (Chapter 2, § 9, Def. 16 (b)) *of a semi-reflexive metrizable t-space E_u is a complete S-space.*

Proof. Since E_u is metrizable it is well known that E'^{β} is complete (Chapter 2, § 9, (7) (b)). Hence in order to prove the theorem we wish to show that E'^{β} is an S-space. Since E_u is semi-reflexive, $E'^{\beta'} = E$ and E'^{β} is a t-space, as follows from ((8) (c)) (Chapter 2, § 9), and also $\beta = \tau(E', E)$. Furthermore, each $\sigma(E', E)$-bounded set is $\tau(E', E)$-precompact as is easy to see (e.g. see [4], Chapitre IV, § 1, N° 6, Exer. 12 (b)). But every $\tau(E', E)$-precompact set, being $\sigma(E', E)$-bounded, is equicontinuous because E is a t-space by hypothesis. Hence $\beta = \tau(E', E)$-precompact, and equicontinuous sets in E' being the same, we have $u = p$ on E. This shows that u coincides with $\sigma(E, E')$ on each equicontinuous set of E because p does so. But $p = u$ being metrizable by hypothesis, $ew^* = p$ by Corollary 1 (Chapter 5, § 2). Hence E'^{β} is a complete S-space.

Corollary 10. *The strong dual of a semi-reflexive Fréchet space is a complete S-space.*

Proof. The corollary follows from Theorem 10 because each Fréchet space is a metrizable t-space.

Corollary 11. *The strong dual of a metrizable Montel* (Chapter 2, § 9, Def. 17) *space is a complete S-space.*

Proof. It is immediate from Theorem 10 because each Montel space is reflexive (hence semi-reflexive) and is a t-space.

Corollary 12. *The strong dual of a Montel space E_u is a complete S-space provided E'^{β} is metrizable.*

Proof. Since the strong dual of a Montel space is a Montel space (Chapter 2, § 9, Prop. 17), the corollary follows from Corollary 11.

6. Countability conditions

In this section we should like to discuss the situation when an l.c. space has a countable fundamental system (Chapter 2, § 6, Def. 10 (b)) of compact or precompact subsets. We shall see that the existence of such a system in an l.c. space E imposes a great restriction on E. The case when there exists a countable fundamental system of bounded sets in an l.c. space has already been treated in other books, e.g. [6]. The case of t-spaces or bornological spaces with a countable fundamental system of compact sets is due to J. Dieudonné [8]; that of quasi-barrelled spaces is due to M. Mahowald and G. Gould [25], and that of complete l.c. spaces with a countable fundamental system of precompact sets is due to T. Husain [14]. Here we discuss only the results of [14].

Proposition 5. *Let E_u be an l.c. space which has a countable fundamental system of u-precompact sets. Then E_u is an S-space.*

Proof. Let ew^* and p be the usual topologies on E'. Clearly, by definition, $ew^* = k(\sigma, \mathscr{C}') = k(p, \mathscr{C}')$ (Chapter 5, § 1, Def. 1 (a)), where $\mathscr{C}'$ is the collection of all $\sigma(E', E)$-closed convex equicontinuous sets of E'. Let $\mathscr{C}$ denote the class of all p-compact convex sets in E'. Then $k(p, \mathscr{C}) \supset k(p, \mathscr{C}')$ by Proposition 2 (Chapter 5, § 1) because each p-compact convex set is $\sigma(E', E)$-compact. Therefore we have

$$k(p, \mathscr{C}) \supset k(p, \mathscr{C}') = ew^* \supset p.$$

But since E_u contains a countable fundamental system of precompact sets, p is metrizable. Therefore by Corollary 1 (Chapter 5, § 2),

$$p = k(p, \mathscr{C}) = ew^*.$$

This proves that E_u is an S-space.

Theorem 11. (*Husain* [14].) *Let E_u be a complete l.c. space which has a countable fundamental system of u-precompact sets. Then E_u is B-complete.*

Proof. By Proposition 5, E_u is an S-space and hence E_u is B-complete due to Corollary 1 (§ 2).

Corollary 13. *Let E be the strict inductive limit of complete l.c. spaces E_n $(n \geqslant 1)$ such that each E_n contains a countable fundamental system of precompact sets. Then E is B-complete.*

Proof. The fact that E is complete follows from Exercise 9 ([3], Chapitre II, § 2, N° 5), and that E contains a countable fundamental system of precompact sets follows from Corollary 2 (Chapter 2, § 6) coupled with the hypothesis. Therefore E is B-complete by Theorem 11.

Corollary 14. *An LF-space E with a defining sequence of Fréchet spaces E_n $(n \geqslant 1)$ is B-complete, provided each E_n contains a countable fundamental system of precompact sets.*

Proof. This is a particular case of Corollary 13.

Note that an LF-space in general is *not* B-complete. An example to this effect is due to A. Grothendieck [11].

Corollary 15. *Let E_n $(n \geqslant 1)$ be a sequence of Fréchet spaces each of which contains a countable fundamental system of precompact sets. If E denotes the direct sum of E_n, then E is B-complete.*

Proof. This is again a particular case of Corollary 13 because the method of constructing a direct sum is a particular case of that of constructing inductive limits.

Corollary 16. *The strong dual of a Montel space E is B-complete provided E contains a countable fundamental system of bounded sets.*

Proof. Clearly E'^{β} is metrizable by hypothesis. Hence by Corollary 12, E is B-complete.

Corollary 17. *A Montel space E_u is B-complete provided E contains a countable fundamental system of bounded sets.*

Proof. By hypothesis, E'^{β} is metrizable and a Montel space (Chapter 2, Prop. 17). Therefore by Corollary 11, $E_u = E'^{\beta'\beta}$ is B-complete.

It would be quite desirable to remove countability conditions in the theorems of this section.

7

LOCALLY CONVEX SPACES WITH THE $B(\mathscr{C})$-PROPERTY

1. The concept of the $B(\mathscr{C})$-property

COMPARING the different notions of hypercomplete, B-complete, B_r-complete, and complete l.c. spaces E dealt with in the preceding chapters, one observes that each of these notions depends upon the coincidence of the ew^*-topology with the $\sigma(E', E)$-topology on some prescribed sets of the dual E' of E. More precisely, let Q be an arbitrary set of the dual E' of an l.c. space E and suppose the following statement is true:

(A′) *Q is $\sigma(E', E)$-closed if and only if it is ew^*-closed.*

Then E is hypercomplete, B-complete, B_r-complete, and complete if (A′) is true for each convex circled set Q, each subspace Q, each dense subspace Q, and each hyperplane Q of E' respectively.

Motivated by the definition of B-complete spaces (Chapter 4, Def. 1 (*a*)), one asks if it is possible to characterize the above-mentioned notions in terms of mappings and range spaces. More precisely, let $\mathscr{C}$ denote a fixed class of l.c. spaces. If a linear continuous and almost open mapping of an l.c. space E onto any l.c. space F in $\mathscr{C}$ is open, then E is said to be a $B(\mathscr{C})$-space. A general problem is to determine the class $\mathscr{C}$ such that a $B(\mathscr{C})$-space is of a prescribed type. In particular, one asks for the class $\mathscr{C}$ such that the class of any one of the types of l.c. spaces described in the preceding paragraph coincides with the class of $B(\mathscr{C})$-spaces. We know (Chapter 4, Def. 1 (*a*)) that E is B-complete if, for each arbitrary l.c. space F, a linear continuous and almost open mapping of E onto F is open. For complete l.c. spaces there exists no $\mathscr{C}$, as shown by Husain [36].

One observes that these two approaches to describing l.c. spaces are, in the obvious sense, dual to each other. However, the advantage of the latter over the first is remarkable because of natural choices of numerous classes $\mathscr{C}$ of l.c. spaces F at our disposal. One sees that the spaces described in the first paragraph are all complete. However, the surprising thing about the $B(\mathscr{C})$-spaces is that one gets out of completeness altogether, as is shown by T. Husain [15] and as is the theme of the present chapter. Moreover, the closed graph and the open mapping theorems, partially on account of which these notions have been investigated, can be carried over to spaces described by the second method as well. The notion of the $B(\mathscr{C})$-property is due to Husain, and all the results of this chapter will be found in his papers [15] and [35]. The concept of $B(\mathscr{C})$-spaces thus generalizes that of B-completeness.

Definition 1. (*a*) Let $\mathscr{C}$ denote a fixed class of l.c. spaces. An l.c. space E is said to have the *$B(\mathscr{C})$-property* or, in short, to be a *$B(\mathscr{C})$-space* if, for each l.c. space F in $\mathscr{C}$, a linear continuous and almost open mapping f of E onto F is open.

(*b*) An l.c. space E is said to be a *$B_r(\mathscr{C})$-space* if, for each l.c. space F in $\mathscr{C}$, a linear continuous one-to-one and almost open mapping of E onto F is open.

PROPOSITION 1. *Every $B(\mathscr{C})$-space is a $B_r(\mathscr{C})$-space.*

Proof. The proof is obvious. But the converse is not likely to be true. An example to this effect is *not* yet known.

PROPOSITION 2. *Let $\mathscr{C}_1$ and $\mathscr{C}_2$ be two classes of l.c. spaces. If $\mathscr{C}_1 \supset \mathscr{C}_2$, then every $B(\mathscr{C}_1)$-space ($B_r(\mathscr{C}_1)$-space) is a $B(\mathscr{C}_2)$-space ($B_r(\mathscr{C}_2)$-space).*

Proof. The proof is obvious.

COROLLARY 1. *Let $B(\mathscr{C})$ denote the class of all $B(\mathscr{C})$-spaces and let $\mathscr{A}$, $\mathscr{I}$, $\mathscr{B}$, $\mathscr{F}$, $\mathscr{B}_n$, $\mathscr{F}_d$ denote the class of* all *l.c., barrelled, Baire, Fréchet, Banach, and finite-dimensional spaces respectively. Then*

$$B(\mathscr{A}) \subset B(\mathscr{I}) \subset B(\mathscr{B}) \subset B(\mathscr{F}) \subset B(\mathscr{B}_n) \subset B(\mathscr{F}_d).$$

Proof. The corollary follows from Proposition 2 by observing that $\mathscr{A} \supset \mathscr{I} \supset \mathscr{B} \supset \mathscr{F} \supset \mathscr{B}_n \supset \mathscr{F}_d$.

Remark. Each of the inclusion relations in Corollary 1 is expected to be strict. So far, only the first and the last inclusion relations are known to be strict ([15], [16]), for which we will give examples later on.

COROLLARY 2. *Every l.c. space E is B-complete (B_r-complete) if and only if E is a $B(\mathscr{A})$-space ($B_r(\mathscr{A})$-space), where $\mathscr{A}$ is defined in Corollary* 1.

Proof. The proof is immediate due to Definition 1 and Definition 1 (Chapter 4).

COROLLARY 3. *Using the notations of Corollary* 1, $B(\mathscr{F}_d) = \mathscr{A}$.

Proof. Clearly $B(\mathscr{F}_d) \subset \mathscr{A}$. For the converse, let E be any l.c. space in $\mathscr{A}$. Let f be a linear continuous and almost open mapping of E onto any finite-dimensional space F. Then by Theorem 1 (*a*) (Chapter 3, § 1), f is open. Hence E is in $B(\mathscr{F}_d)$ by Definition 1, and therefore $\mathscr{A} = B(\mathscr{F}_d)$.

Example. Now to show that $B(\mathscr{B}_n) \subset B(\mathscr{F}_d)$ is strict, consider the example given in Proposition 1 (Chapter 4). E_ω (the point set of an infinite-dimensional Banach space E_u endowed with the finest locally convex topology ω) is clearly in $B(\mathscr{F}_d)$ due to Corollary 3. However, E_ω does not belong to $B(\mathscr{B}_n)$, because the identity mapping $E_\omega \to E_u$ is linear continuous and almost open but *not* open (Chapter 4, § 1, Prop. 1). Actually E_ω does *not* belong to any class of l.c. spaces considered in Corollary 1 except to that of $B(\mathscr{F}_d)$.

In view of the characterization of a B-complete space (Chapter 4, § 1, Theorem 1) in terms of subspaces of its dual, it is natural to ask for a similar characterization of $B(\mathscr{C})$-spaces. In the sequel, we shall give a characterization of $B(\mathscr{C})$-spaces when $\mathscr{C} = \mathscr{I}$.

2. The closed graph theorem for $B_r(\mathscr{C})$-spaces

Let us assume that a fixed class $\mathscr{C}$ of l.c. spaces satisfies the following condition:

(B′) *Let E and F be two l.c. spaces and let f be a linear continuous and almost open mapping of E onto F. If E is in $\mathscr{C}$ then F is also in $\mathscr{C}$.*

Remark. A consequence of the assumption (B′) on the class $\mathscr{C}$ is that the quotient of each l.c. space in $\mathscr{C}$ is also in $\mathscr{C}$.

THEOREM 1. (*Husain* [15].) *Let a fixed class $\mathscr{C}$ of l.c. spaces satisfy* (B′). *Let F_v be an l.c. space and E_u in $\mathscr{C}$ a $B_r(\mathscr{C})$-space. Let f be a linear mapping of F_v into E_u, the graph of which is closed in $F \times E$. If f is almost continuous, then f is continuous.*

Proof. In view of Proposition 18 (Chapter 2, § 9), it is sufficient to prove that $f\colon F_\sigma \to E_u$ is continuous because f is almost continuous, where $\sigma = \sigma(F, F')$. Let Q denote the set of all linear functionals $x' \in E'_u$ such that $\langle f(y), x' \rangle$ is continuous in $y \in F_v$. According to Lemma 2 (Chapter 4, § 5), Q is a dense subspace of E' because the graph of f is closed by hypothesis; and due to Lemma 3 (Chapter 4, § 5), Q is almost closed in E' because f is almost continuous.

Now the density of Q in E'^σ implies $Q' = E$. Let w denote the $\mathfrak{S}$-topology on E with sets $(Q \cap U^0)^0$ as a fundamental system of neighbourhoods of 0, where U runs over all convex circled u-neighbourhoods of 0 in E. The fact that w is an l.c. topology is easy to verify. To show that w is Hausdorff, let $x \in (Q \cap U^0)^0$ for each U. Then $\langle x, Q \cap U^0 \rangle \leqslant 1$ for every U. Whence it follows that $\langle x, Q \rangle = 0$. Since Q is dense in E', $x = 0$. Further, each $Q \cap U^0$ is convex circled and $\sigma(E', E)$-compact, because $Q \cap U^0$ is a $\sigma(E', E)$-closed (since Q is almost closed) subset of a $\sigma(E', E)$-compact set U^0. Hence by Theorem 6 (Chapter 2, § 9), w is an $\mathfrak{S}$-topology compatible with the duality between E and Q, i.e. $E'_w = Q$. Since $Q \cap U^0 \subset U^0$, therefore $W = (Q \cap U^0)^0 \supset U$ for each W. Hence $u \supset w$. Now we show that $w(u) = w$. Clearly $w(u) \supset w$ and $w(u)$ is Hausdorff (Chapter 4, § 2, Theorem 2), because $u \supset w$. Let V be a convex circled w-closed $w(u)$-neighbourhood of 0. Then $V = wU$ where wU is a convex circled w-closed u-neighbourhood of 0, and

$$V^Q = (wU)^Q = U^Q = U^0 \cap Q.$$

But $V = V^{QE}$, because V is a convex circled w-closed neighbourhood of 0. Therefore,

$$V = V^{QE} = (U^0 \cap Q)^0$$

which is a w-neighbourhood of 0 by definition. Hence $w(u) = w$. Now $u \supset w$ and $w(u) = w$ together mean that the identity mapping: $E_u \to E_w$ is continuous and almost open. Hence E_w is in $\mathscr{C}$ by assumption on $\mathscr{C}$ because E_u is in $\mathscr{C}$ by hypothesis. Now the fact that E_u is a $B_r(\mathscr{C})$-space by hypothesis implies $u = w$. This shows that $Q = E'_w = E'_u$. In other words, f is weakly continuous, and thus the proof is completed.

Remark. Since an l.c. space is B_r-complete if and only if it is a $B_r(\mathscr{A})$-space, and since the assumption (B′) is trivially true for the class $\mathscr{A}$, Theorem 6 (Chapter 4, § 5) is an immediate consequence of Theorem 1.

COROLLARY 4. *Let F_v be any l.c. space and E_u a quasi-barrelled and also a $B_r(\mathscr{Q})$-space, where $\mathscr{Q}$ is the class of all quasi-barrelled l.c. spaces. Let f be a linear mapping of F_v into E_u, the graph of which is closed in $F \times E$. If f is almost continuous, then f is continuous.*

Proof. This is immediate from Theorem 1 by observing that the assumption (B′) made on the class $\mathscr{C}$ in Theorem 1 is true for the class $\mathscr{Q}$, in view of (5) (*d*) (Chapter 2, § 6).

COROLLARY 5. *Let F_v be an l.c. space and E_u a t-space and also a $B_r(\mathscr{I})$-space. Let f be a linear mapping of F_v into E_u, the graph of which is closed in $F \times E$. If f is almost continuous, then f is continuous.*

Proof. In view of Proposition 5 (Chapter 2, § 6), the assumption (B′) (see § 2) on $\mathscr{I}$, the class of all t-spaces, is satisfied; and therefore the corollary follows from Theorem 1.

3. The $B_r(\mathscr{I})$- and the $B(\mathscr{I})$-spaces

We should like to specialize the $B(\mathscr{C})$-spaces in order to show, among other results, that in the case of $\mathscr{C} = \mathscr{I}$, a $B(\mathscr{I})$-space need *not* be complete and therefore *not* B-complete.

Definition 2. Let E' be the dual of an l.c. space E_u. A subspace Q of E', when Q is endowed with the relative $\sigma(E', E)$-topology, is said to be *boundedly complete* if the following conditions are simultaneously satisfied:

(a') For each u-neighbourhood U of 0 in E_u, $Q \cap U^0$ is closed in E'.
(b') Every bounded set of Q is equicontinuous.

PROPOSITION 3. *Every boundedly complete subspace of the dual E' of an l.c. space E_u is almost closed.*

Proof. In view of Definition 2 (Chapter 4, § 1), the proposition follows from (a') of Definition 2.

PROPOSITION 4. *Every boundedly complete subspace Q of the dual E' of an l.c. space E_u is quasi-complete.*

Proof. Let B be a bounded subset of Q. Since Q is boundedly complete, B is equicontinuous due to (b') of Definition 2. But this means that B as a subset of E' is equicontinuous. Hence by Proposition 14 (Chapter 2, § 9), there exists a u-neighbourhood U of 0 such that $B \subset U^0$. But also $B \subset Q$ implies $B \subset Q \cap U^0$ which is compact due to (a') of Definition 2 and the fact that U^0 is compact in E'. Therefore B is relatively compact. Hence Q is quasi-complete.

THEOREM 2. *A necessary and sufficient condition for an l.c. space E_u to be a $B_r(\mathscr{I})$-space is that each dense boundedly complete subspace of the dual E' of E coincides with E'.*

Proof. For the 'necessary' part, assume that E_u is a $B_r(\mathscr{I})$-space. Let Q be a dense boundedly complete subspace of E', where Q is endowed with the relative $\sigma(E', E)$-topology. The density of Q implies $Q' = E$. Let v denote the $\mathfrak{S}$-topology on E, where $\mathfrak{S}$ consists of all bounded sets of Q. In other words, v has the sets B^0 as a fundamental system of neighbourhoods of 0, when B runs over all bounded sets of Q. Clearly v is the strong topology on E and therefore $v \supset \tau(E, Q)$. On the other hand, according to (a') of Definition 2, for each convex circled u-neighbourhood U of 0, $Q \cap U^0$ is a convex circled and $\sigma(E', E)$-compact set of Q, and therefore $(Q \cap U^0)^0$ is a $\tau(E, Q)$-neighbourhood of 0. But according to (b') of Definition 2, for each bounded set B of Q, $B \subset Q \cap U^0$ implies $B^0 \supset (Q \cap U^0)^0$. This shows that $v \subset \tau(E, Q)$. Combining the two inequalities between v and $\tau(E, Q)$, we have $v = \tau(E, Q)$. Since Q is quasi-complete by Proposition 4, it is easy to verify that E_v is a t-space. Now we have two locally convex topologies u and v on E. We show that $u \supset v$. Let V be a v-neighbourhood of 0. Then there exists a bounded set B in Q such that $V \supset B^0$. But Q being boundedly complete, (b') of Definition 2 implies $B \subset U^0$ for some u-neighbourhood U of 0 in E. But then

$$V \supset B^0 \supset U^{00} \supset U$$

implies $u \supset v$. In other words, the identity mapping i of E_u onto a t-space E_v is continuous. Clearly i is one-to-one, and due to Proposition 6 (b) (Chapter 2, § 6), it is almost open. Hence i is open because E_u is a $B_r(\mathscr{I})$-space by assumption. Therefore $u = v$ implies $Q = E'_v = E'_u$.

For the 'sufficient' part, assume that every dense boundedly complete subspace of E' coincides with E'. Let F be a t-space and f a linear continuous one-to-one mapping of E_u onto F. Due to f being one-to-one and 'onto', it is sufficient to assume that E_u is endowed with another l.c. topology v such that $u \supset v$ and E_v is a t-space. Obviously $Q = E'_v$ is dense in E'_u because the identity mapping is one-to-one.

Now we show that Q is a boundedly complete subspace of E'_u. By Proposition 6 (b) (Chapter 2, § 6), the identity mapping: $E_u \to E_v$ is almost open, i.e. for each u-neighbourhood U of 0, vU is a v-neighbourhood of 0 and hence $(vU)^Q$ is $\sigma(Q, E_v)$-compact and therefore $\sigma(E'_u, E_u)$-compact because the mapping: $Q \to E'$ is weakly continuous. But then

$$(vU)^Q = U^Q = U^0 \cap Q$$

proves that $U^0 \cap Q$, being $\sigma(E'_u, E_u)$-compact, is $\sigma(E'_u, E_u)$-closed, because $\sigma(E', E)$ is a Hausdorff topology. This shows that part (a') of Definition 2 is satisfied. For (b'), let B be a circled bounded set of Q. Then B^0 is a v-barrel and therefore a v-neighbourhood of 0 because E_v is a t-space. But $u \supset v$ implies that there exists a u-neighbourhood U of 0 such that $B^0 \supset U$. Hence $B \subset B^{00} \subset U^0$ establishes that Q is boundedly complete and therefore by assumption, $Q = E'_v = E'_u$. This shows that u is compatible with the duality between E_v and Q. But v being the Mackey topology (because E_v is a t-space), it follows that $v \supset u$. Since $u \supset v$ by assumption, therefore $u = v$. This proves that f is open. Hence E_u is a $B_r(\mathscr{I})$-space.

A similar characterization for $B(\mathscr{I})$-spaces is as follows:

THEOREM 3. *A necessary and sufficient condition for an l.c. space E_u to be a $B(\mathscr{I})$-space is that each boundedly complete subspace of the dual E' of E is $\sigma(E', E)$-closed.*

Proof. For the 'necessary' part, assume E_u is a $B(\mathscr{I})$-space. Let Q be a boundedly complete subspace of E', when Q is endowed with the relative $\sigma(E', E)$-topology. Clearly $Q' = E/Q^0$. Let β denote the strong topology on Q'. As shown in Theorem 2, Q'^{β} is a t-space because

Q is boundedly complete, and $\beta = \tau(Q', Q)$. First of all we show that the mapping $\varphi: E_u \to Q'^{\beta}$ is continuous. For an arbitrary β-neighbourhood V of 0 in Q'^{β}, there exists a bounded set B in Q such that $B^{Q'} \subset V$. But the mapping: $Q_\sigma \to E'^{\sigma}$ being continuous implies $V \supset B^{Q'} = \varphi(B^0)$, where B^0 is the polar of B in E. Since every bounded set in Q is equicontinuous (because Q is boundedly complete), there exists a u-neighbourhood U of 0 in E_u such that $B^0 \supset U$. Whence

$$\varphi^{-1}(V) \supset \varphi^{-1}(\varphi(B^0)) \supset \varphi^{-1}(\varphi(U)) \supset U.$$

This shows that φ is continuous and therefore φ is open because E_u is a $B(\mathscr{I})$-space by assumption. In other words, $Q = Q^{00}$ implies Q is $\sigma(E', E)$-closed.

For the 'sufficient' part, assume that every boundedly complete subspace of E' is $\sigma(E', E)$-closed. Let F be a t-space, and let f be a linear continuous mapping of E onto F. Then the transpose mapping $f': F' \to E'$ is a homeomorphism into. Let F' be identified with its image $f'(F')$ in E'. Since F is a t-space, F' is boundedly complete, as can be shown in exactly the same manner as in Theorem 2. Therefore by assumption, F' is closed in E'. From this it follows easily that f is open (Chapter 2, § 9, Prop. 19).

4. $B(\mathscr{L}\mathscr{I})$-spaces and $B(\mathscr{I})$-spaces

PROPOSITION 5. *Let E be a $B(\mathscr{I})$-space ($B_r(\mathscr{I})$-space) and M a closed subspace of E. Then E/M is a $B(\mathscr{I})$-space ($B_r(\mathscr{I})$-space).*

Proof. Obviously $(E/M)' = M^0$. Let Q be a boundedly complete subspace of M^0. Since Q can be regarded as a subspace of E' and since $\sigma(E', E)$ coincides with $\sigma(M^0, E/M)$ on M^0, therefore Q can be regarded as a boundedly complete subspace of E'. But then it implies that Q is $\sigma(E', E)$-closed in E', because E is a $B(\mathscr{I})$-space by hypothesis. Hence it follows that Q is $\sigma(M^0, E/M)$-closed in M^0. The statement about $B_r(\mathscr{I})$-spaces follows in the same way.

Definition 3. Let $\mathscr{C}$ be a fixed class of l.c. spaces. Let F_n ($n \geqslant 1$) be a strictly increasing sequence of l.c. spaces such that F_n is in $\mathscr{C}$ for each n. Assume that F_n is a closed subspace of F_{n+1} for each n. Then $\mathscr{L}\mathscr{C}$ denotes the class of all strict inductive limits of sequences $\{F_n\}$, F_n in $\mathscr{C}$. An l.c. space E which satisfies the $B(\mathscr{L}\mathscr{C})$-property is called a $B(\mathscr{L}\mathscr{C})$-space.

In view of ((2) (*a*)) (Chapter 2, § 6), $\mathscr{I} = \mathscr{L}\mathscr{I}$, where $\mathscr{I}$ is the class of all t-spaces. However, for a subclass, $\mathscr{F}$ or $\mathscr{B}$ of $\mathscr{I}$, $\mathscr{L}\mathscr{F} \supset \mathscr{F}$ and $\mathscr{L}\mathscr{B} \supset \mathscr{B}$. In the following theorem, $\mathscr{I}'$ denotes any fixed subclass of $\mathscr{I}$, and $\mathscr{L}\mathscr{I}'$ the class of all strict inductive limits of sequences of spaces in $\mathscr{I}'$.

THEOREM 4. *An l.c. space E is a $B(\mathscr{L}\mathscr{I}')$-space if each closed subspace of E is a $B(\mathscr{I}')$-space.*

Proof. Let F be an l.c. space in $\mathscr{L}\mathscr{I}'$ with its defining sequence F_n $(n \geqslant 1)$, F_n in $\mathscr{I}'$. Let f be a linear continuous and almost open mapping of E onto F. For each n, $f^{-1}(F_n)$ is a closed subspace of E because f is continuous and F_n is closed. Let f_n denote the restriction of f on $f^{-1}(F_n)$. Since F_n is a t-space, f_n is a linear and almost open mapping of $f^{-1}(F_n)$ onto F_n. Further, for each neighbourhood U of 0 in F, $U \cap F_n$ is a neighbourhood of 0 in F_n and therefore

$$f_n^{-1}(U \cap F_n) = f^{-1}(U) \cap f^{-1}(F_n)$$

is a neighbourhood of 0 in $f^{-1}(F_n)$ because f is continuous. This implies f_n is continuous and hence, by hypothesis, f_n is open. Hence for each neighbourhood V of 0 in E, $V \cap f^{-1}(F_n)$ being a neighbourhood of 0 in $f^{-1}(F_n)$, it follows that

$$f_n(V \cap f^{-1}(F_n)) = f(V) \cap F_n$$

is a neighbourhood of 0 in F_n for each n. But this implies, by the definition of inductive limits, that $f(V)$ is a neighbourhood of 0 in F. In other words, f is open. This completes the proof.

COROLLARY 6. *Let E be an l.c. space and F the strict inductive limit of Baire spaces (or let F be an LF-space). Then a linear and continuous mapping of E onto F is open provided each closed subspace of E is a $B(\mathscr{B})$-space (or a $B(\mathscr{F})$-space).*

Proof. This is an immediate consequence of Theorem 4.

COROLLARY 7. *Let E be B-complete and F the strict inductive limit of Baire spaces. Then a linear and continuous mapping of E onto F is open.*

Proof. In view of Corollary 2, due to Proposition 4 (Chapter 4, § 1), each closed subspace of a B-complete l.c. space, being B-complete, is a $B(\mathscr{I})$-space and therefore a $B(\mathscr{B})$-space. Hence the corollary follows from Corollary 6.

A particular case of Corollary 7 is the following:

COROLLARY 8. *Let E be a Fréchet space and F an LF-space. Then a linear and continuous mapping of E onto F is open.*

A more general case of Corollary 7, when E is the inductive limit of B-complete spaces, is Theorem 9 (a) (Chapter 4, § 5).

A few more general cases of Corollary 8 are Theorems 10, 11, 12 (Chapter 3, § 3).

5. The closed-graph theorem for $B_r(\mathscr{I})$-spaces

Before we prove the main result of this section, we show that Corollary 5 is actually another version of Theorem 6 (Chapter 4, § 5).

THEOREM 5. *Every t-space E which is a $B_r(\mathscr{I})$-space is B_r-complete.*

Proof. Let Q be an almost closed dense subspace of the dual E' of E. Let B be a bounded subset of Q, when Q is endowed with the relative $\sigma(E', E)$-topology. Then B is $\sigma(E', E)$-bounded in E' and hence equicontinuous because E is a t-space. This shows that Q is boundedly complete. But then E being a $B_r(\mathscr{I})$-space it follows that Q is $\sigma(E', E)$-closed. Hence $Q = E'$, and therefore E is B_r-complete.

The following main theorem of this section is a particular case of a very general theorem for which the reader is referred to Husain's paper [35].

THEOREM 6. *Let E_u be a t-space and F_v a $B_r(\mathscr{I})$-space. Let f be a linear mapping of E into F so that the graph of f is closed in $E \times F$. If f is almost open, then f is continuous.*

Proof. Let $\{V\}$ denote a fundamental system of closed, convex, and circled neighbourhoods of 0 in F_v. For each V in $\{V\}$, let

$$\overset{*}{V} = \overline{f(\overline{f^{-1}(V)})}.$$

Then it is easy to verify that $\{\overset{*}{V}\}$ forms a filterbase for convex, closed, and circled neighbourhoods of 0 in F under a locally convex topology w. We show that w is Hausdorff. Let $y \in \overset{*}{V}$, for each $\overset{*}{V}$ in $\{\overset{*}{V}\}$. Then $y \in \frac{1}{2}\overset{*}{V}$. Hence there exists $x_1 \in \overline{f^{-1}(\frac{1}{2}V)}$ so that $f(x_1) \in y + \frac{1}{2}V$. But $\overline{f^{-1}(\frac{1}{2}V)} \subset f^{-1}(\frac{1}{2}V) + U$, where U is any arbitrary neighbourhood of 0 in E. Hence $x_1 \in f^{-1}(\frac{1}{2}V) + U$. That means there exists $x_2 \in U$ so that $x_1 - x_2 \in f^{-1}(\frac{1}{2}V)$ and hence $f(x_2) \in f(x_1) + \frac{1}{2}V$. But this shows that $f(x_2) \in y + V$. In other words, $G \cap (U \times (y+V)) \neq \emptyset$, where G is the graph of f in $E \times F$. Since G is closed by hypothesis, $(0, y) \in G$ and hence $y = 0$.

Further, for each $\overset{*}{V}$, $\overset{*}{V} \supset V$ so $v \supset w$. Hence $v \supset w(v) \supset w$ by § 2 (Chapter 4). We wish to show that $w(v) = w$. For this it is sufficient to show that if $y \in \overset{*}{V}$ then $y \in wV$. For each W in $\{V\}$,

$$(y + \tfrac{1}{2}W) \cap f(\overline{f^{-1}(V)}) \neq \emptyset.$$

Hence there exists $x_1 \in \overline{f^{-1}(V)}$ so that $f(x_1) \in y+\frac{1}{2}W$. But $\overline{f^{-1}(\frac{1}{2}W)}$, being a barrel in E, is a neighbourhood of 0 (because E is a t-space). Hence $x_1 \in f^{-1}(V)+\overline{f^{-1}(\frac{1}{2}W)}$. Therefore there exists $x_2 \in f^{-1}(V)$ such that $x_1-x_2 \in \overline{f^{-1}(\frac{1}{2}W)}$ and hence $f(x_1)-f(x_2) \in \frac{1}{2}\overset{*}{W}$. But then

$$f(x_2) \in y+\tfrac{1}{2}W+\tfrac{1}{2}\overset{*}{W}$$

(because $\frac{1}{2}\overset{*}{W}$ is circled and $f(x_1) \in y+\frac{1}{2}W$). Hence $f(x_2) \in y+\overset{*}{W}$, because $W \subset \overset{*}{W}$ and $\overset{*}{W}$ is convex, and therefore $V \cap (y+\overset{*}{W}) \neq \emptyset$ for each $\overset{*}{W}$. This proves that y is a limit point of V under w, or in other words, $y \in wV$. Thus the identity mapping: $F_v \to F_w$ is continuous and almost open (Prop. 9, Chapter 4).

Now consider the mapping $f\colon E_u \to F_w$. For each $\overset{*}{V}$ in $\{\overset{*}{V}\}$,

$$f^{-1}(\overset{*}{V}) = f^{-1}(\overline{f(\overline{f^{-1}(V)})}) \supset \overline{f^{-1}(V)}.$$

Since $\overline{f^{-1}(V)}$ is a barrel and therefore a neighbourhood of 0 in E_u, it shows that $f\colon E_u \to F_w$ is continuous. Further, for each u-neighbourhood U of 0 in E,

$$wf(U) \supset vf(U)$$

because $v \supset w$. Hence

$$wf(U) \supset w(vf(U)).$$

Since $f\colon E_u \to F_v$ is almost open by hypothesis, $vf(U)$ is a v-neighbourhood of 0 in F. But then $w(vf(U))$ is a w-neighbourhood of 0 in F, because $w(v) = w$. Therefore $f\colon E_u \to F_w$ is almost open.

Now $f\colon E_u \to F_w$ being a continuous and almost open mapping of a t-space E_u into an l.c. space F_w implies F_w is a t-space by Proposition 5 (Chapter 2). Since F_v is a $B_r(\mathscr{I})$-space by hypothesis, the identity mapping: $F_v \to F_w$, being continuous and almost open, is open. Hence $v = w$. Since $f\colon E_u \to F_w$ has been proved to be continuous, it follows that $f\colon E_u \to F_v$ is continuous. This completes the proof of the theorem.

It is not yet known whether or not almost openness in Theorem 6 can be dropped.

Note. The consideration of the topology w in the above theorem is due to A. and W. Robertson [31] in connexion with another closed graph theorem (Chapter 4, Theorem 8 (b)).

THEOREM 7. *Let E_u be a t-space and F_v a $B_r(\mathscr{I})$-space. Then a linear almost continuous mapping g of F_v onto E_u with the closed graph is open.*

Proof. Since the graph of g is closed, $g^{-1}(0)$ is a closed linear subspace of F_v (Chapter 3, Lemma 3). Since the quotient space of a $B_r(\mathscr{I})$-space is a $B_r(\mathscr{I})$-space (Prop. 5), in view of (1) (Chapter 3, § 2) it can be assumed that g is 1:1. Hence g^{-1}: $E_u \to F_v$ exists and is almost open because g is almost continuous. Therefore Theorem 6 applies and hence g^{-1} is continuous. This shows that g is open.

Another characterization of $B_r(\mathscr{I})$-spaces is as follows:

THEOREM 8. *Let E be a t-space. Then E is a $B_r(\mathscr{I})$-space if and only if, for each t-space F, a linear mapping f of E onto F with the closed graph is open.*

Proof. Assume E is a $B_r(\mathscr{I})$-space. Then according to Theorem 5 and Corollary 4 (Chapter 4), f is open. On the other hand, if f is a continuous mapping of E onto any t-space F, then the graph of f is closed in $E \times F$ (Chapter 3, § 1, Prop. 1), and therefore f is open by assumption. This shows that E is a $B(\mathscr{I})$-space and *a fortiori* a $B_r(\mathscr{I})$-space by Proposition 1.

6. Examples and counter-examples

LEMMA 1. *A sufficient condition for an l.c. space E_u to be a $B(\mathscr{I})$-space ($B_r(\mathscr{I})$-space) is that each quasi-complete (dense and quasi-complete) subspace of E' is closed (coincides with E').*

Proof. In view of Proposition 4, the lemma follows from Theorems 2 and 3.

PROPOSITION 6. *Let E_u be a metrizable l.c. space and E' its dual. Then E', endowed with any locally convex topology finer than $\sigma(E', E)$ and coarser than $\tau(E', E)$, is a $B_r(\mathscr{I})$-space.*

Proof. Let v be a locally convex topology such that

$$\sigma(E', E) \subset v \subset \tau(E', E).$$

Clearly $E'^{v'} = E$. Let Q be a quasi-complete subspace of E. Then Q is also quasi-complete under $\tau(E, E') = u$. But then Q is a Fréchet space (because E_u is metrizable) and therefore u-closed. Q being convex, it is $\sigma(E, E')$-closed by Proposition 15 (Chapter 2, § 9). Hence by Lemma 1, E'^v is a $B(\mathscr{I})$-space and *a fortiori* a $B_r(\mathscr{I})$-space.

Remark. Comparing Proposition 6 with Proposition 7 (Chapter 4, § 1), one observes that the $B(\mathscr{I})$-property of a space is much weaker than B-completeness. Proposition 6 provides many examples of $B(\mathscr{I})$-spaces which are not complete and therefore *not* B-complete, as will be seen later on.

PROPOSITION 7. *Let E_u be an l.c. space and assume that every subspace Q of E' is finite-dimensional whenever each bounded subset of Q is finite-dimensional. Then E_σ (where $\sigma = \sigma(E, E')$) is a $B(\mathscr{I})$-space.*

Proof. Let Q be a boundedly complete (§ 3, Def. 2) subspace of E', endowed with the relative $\sigma(E', E)$-topology. Let B be any bounded set in Q. Since B is equicontinuous by the assumption that Q is boundedly complete, and since every equicontinuous set in E' is finite-dimensional when E is endowed with $\sigma(E, E')$, therefore B is finite-dimensional. But this shows that Q itself must be finite-dimensional by hypothesis. Hence Q must be closed by Corollary 3 (Chapter 3, § 1). This, by Theorem 3, proves that E_σ is a $B(\mathscr{J})$-space.

Proposition 8. *Let E_u be a semi-reflexive space with a countable fundamental system of u-bounded sets. Then E_v is a $B(\mathscr{J})$-space for any locally convex topology v such that $\sigma(E, E') \subset v \subset \tau(E, E')$.*

Proof. Semi-reflexivity implies $E'^{\beta'} = E$ and $\beta = \tau(E', E)$. By hypothesis E'^{β} is a metrizable l.c. space. Hence E_v is a $B(\mathscr{J})$-space due to Proposition 6.

Example 1. Let E_n $(n \geqslant 1)$ be a sequence of Fréchet spaces. Then $E = \prod_{n=1}^{\infty} E_n$ is a Fréchet space and therefore $E' = \sum_{n=1}^{\infty} E'_n$ is a $B(\mathscr{J})$-space for all l.c. topologies finer than $\sigma(E', E)$ and coarser than the direct sum topology, due to Proposition 6.

Example 2. Let E_n $(n \geqslant 1)$ be a sequence of l.c. spaces such that $E_n'^{\tau_n}$ is metrizable, where $\tau_n = \tau(E'_n, E_n)$. Then $E'^{\tau} = \prod_{n=1}^{\infty} E_n'^{\tau_n}$ being metrizable, $E = \sum_{n=1}^{\infty} E_n$ is a $B(\mathscr{J})$-space for all l.c. topologies finer than $\sigma(E, E')$ and coarser than the direct sum topology, due to Proposition 6.

(1) Let E_u be the direct sum of a countable number of reflexive Banach spaces E_n $(n \geqslant 1)$ where u is any l.c. topology coarser than the direct sum topology and finer than $\sigma(E, E')$. Let F be a t-space. Then a linear continuous mapping of E onto F is open. For, $E_n'^{\beta_n} = E_n'^{\tau_n}$ is a Banach space and therefore $E'^{\tau} = \prod_{n=1}^{\infty} E_n'^{\tau_n}$ is a Fréchet space. Hence, by Proposition 6, E_u is a $B(\mathscr{J})$-space.

Example 3. Let $E_u = L_1(N)$, the space of all real sequences x_n $(n \geqslant 1)$ such that $\sum_{n=1}^{\infty} |x_n| < \infty$, where u is the metric topology induced from R^N (the countable product of the reals). E_u is dense in R^N and therefore *not* complete, and *a fortiori not* B-complete. But $E'_u = R^{(N)}$, the space of all finite sequences where $\tau(E', E)$ is the norm topology defined by the norm: $\|x\| = \sup |x_n|$. Furthermore, it is known that $E_u'^{\tau'} = E$

and $\sigma(E, E') = u = \tau(E, E')$, therefore E_u is a $B(\mathscr{I})$-space by Proposition 6.

Example 4. Let F_v denote $R^{(N)}$, where $v = \tau(F, F')$, $F' = L_1(N)$. v is the norm topology defined in Example 3. Exactly by the same argument as used in Example 3, F_v is a $B(\mathscr{I})$-space but *not* complete because otherwise it would be a Banach space which is not true.

(2) A linear mapping of a $B(\mathscr{I})$-space (or of a t-space) onto a $B(\mathscr{I})$-space with the closed graph need *not* be open.

For this, let ω denote the finest locally convex topology on $F = R^{(N)}$ and let F_v denote the same TVS as discussed in Example 4. Clearly $F'_\omega = R^N$ which includes $L_1(N)$ as a proper subset. Therefore ω is strictly finer than v. It is known that F_ω is a t-space (Chapter 2, § 6, (2) (*e*)). It is also known (Chapter 6, Theorem 2 and Corollary 1) that F_ω is B-complete and therefore a $B(\mathscr{I})$-space. Now $\omega \supset v$ implies that the graph of the identity mapping: $F_\omega \to F_v$ is closed. But the identity mapping is *not* open.

Moreover, $R^{(N)}$, with the finest locally convex topology, supplies an example of a *non-metrizable* $B(\mathscr{I})$-space.

The following example which is due to Pták [28] shows:

(3) That a linear and continuous mapping of a B-complete space onto another B-complete space need *not* be open.

For this consider the example $E_u = L_1(N)$ with the norm topology defined by the norm: $\|x\| = \sum_{n=1}^{\infty} |x_n| < \infty$. Let v denote another locally convex topology on E defined by the semi-norms:

$$p_\alpha(x) = \sum_{n=1}^{\infty} \alpha_n |x_n|$$

where $\alpha \in A$, the set of all real monotonically decreasing sequences $\alpha = (\alpha_n)$ such that $0 < \alpha_n \leqslant 1$, $\lim_{n\to\infty} \alpha_n = 0$. Since $p_\alpha(x) \leqslant \|x\|$ for each x, $u \supset v$. Further, $E'_u = L_\infty(N)$ and $E'_v = C_0(N)$, the space of all real convergent sequences which converge to 0. Clearly $C_0(N)$ is a proper subspace of $L_\infty(N)$ and therefore $u \supset v$ and $u \neq v$. In other words, the identity mapping: $E_u \to E_v$ is continuous but *not* open. On the other hand, $C_0(N)$ is a Banach space in the sup norm induced from $L_\infty(N)$ and has $E = L_1(N)$ as its dual. It is easy to see that v can be identified with the $\mathfrak{S}$-topology on E, where $\mathfrak{S}$ consists of all convex compact subsets of $C_0(N)$. Hence E_v is B-complete due to Proposition 7 (Chapter 4, § 1). But E_u is B-complete due to Proposition 3 (Chapter 4, § 1) because E_u is a Banach space. This establishes (3).

(4) (*a*) A linear and continuous mapping of a t-space and/or $B(\mathscr{I})$-space onto a $B(\mathscr{I})$-space is *not* necessarily open.

(*b*) A linear mapping of a $B(\mathscr{I})$-space into a t-space and/or $B(\mathscr{I})$-space with the closed graph is *not* necessarily continuous.

For this, consider $E_u = L_1(N)$ which is a Banach space with the norm topology defined in (3). Hence E_u is a t-space and also a $B(\mathscr{I})$-space. Let v be the metric topology on E, induced from R^N. Then E_v is a $B(\mathscr{I})$-space by Example 3. But it is well known that $u \supset v$ but $u \neq v$, because $E'_u = L_\infty(N) \neq R^{(N)} = E'_v$. Thus the identity mapping i: $E_u \to E_v$ answers (*a*) while the identity mapping j: $E_v \to E_u$ answers (*b*), because the graph of j is the same as that of i and therefore closed because the graph of i is closed due to continuity of i.

(5) A $B(\mathscr{F})$-space is *not* necessarily a Fréchet space. In other words, an l.c. space E for which the statement (A) (Chapter 3, § 1) is true for each $F \in \mathscr{F}$, is *not* necessarily a Fréchet space.

For this, consider $R^{(N)}$, with the finest l.c. topology. It is a $B(\mathscr{F})$-space but not a Fréchet space.

(6) (*a*) A linear continuous mapping of a B-complete space onto a $B(\mathscr{I})$-space is *not* necessarily open.

(*b*) A linear mapping of a $B(\mathscr{I})$-space into a B-complete space with the closed graph is *not* necessarily continuous.

These follow from the example given in (2).

Exercises

(1) Let $\mathscr{C}$ be a class of l.c. spaces. An l.c. space E is said to be a *$C(\mathscr{C})$-space* if, for each $F \in \mathscr{C}$, a linear almost continuous mapping of F into E with the closed graph is continuous.

(*a*) Every $C(\mathscr{C})$-space is a $B_r(\mathscr{C})$-space.

(*b*) An l.c. space is a $B_r(\mathscr{A})$-space if and only if it is a $C(\mathscr{A})$-space.

(*c*) Is (*b*) true when $\mathscr{A}$ is replaced by $\mathscr{I}$?

(2) Let E_u be a t-space and F_v a Mackey space which is also a $B_r(\mathscr{I})$-space. Let f be a linear mapping of E_u into F_v the graph of which is closed. Let w be the same topology defined on F as in Theorem 6, § 5. If the transpose mapping f': $F'_w \to E'_u$ has the inverse which is continuous with respect to weak* topologies on F'_w and E'_u respectively, then f is continuous. (See [38].)

8

HISTORICAL NOTES

I DO not wish to give an accurate chronological account of the entire material covered in this book. For a historical background of topological vector spaces, to mention a few, the Historical Notes by N. Bourbaki [4], a colloquium lecture by J. Dieudonné [7], and the *Readers' Guide* by M. M. Day ([6], chapter viii) for normed spaces are sufficient to enthuse anybody and enlighten everybody.

Since Banach's monograph [1] in 1932, which is undoubtedly an inexhaustible source of inspiration for functional analysts and in which perhaps for the first time the so-called Banach's open or interior mapping (or homomorphism) theorem and closed graph theorem for *F*-spaces appeared, several books on topological vector spaces have been published.

To mention a few, the first volume of the book *Espaces Vectoriels Topologiques*, chapters i–ii, by N. Bourbaki [3] appeared in 1953 in the same traditional form which has become a sort of vogue of the day. The second volume [4] appeared in 1955. It may be pointed out that the standardized Bourbakian techniques and terminology have been freely used throughout in this book.

Right at the heel of Bourbaki's first volume, in 1954 *Espaces Vectoriels Topologiques* by A. Grothendieck [13] contributed a great deal of information about the subject even though its contents partly overlapped with [3] and [4]. After that another monograph, *Normed Linear Spaces* by M. M. Day [6], was published in 1958 containing a compendium of results on normed spaces.

Lastly and recently, in 1960, a very good publication of G. Köthe, *Topologische Lineare Räume* I [22], appeared which will undoubtedly remain a guiding beacon to many an adventurer in this field for years to come. The second volume is expected.

The present monograph is necessitated by the fact that though, during the span of the last decade, considerable progress in understanding the open mapping, the closed graph, and the Krein–Šmulian theorems has been made, yet nowhere in the mathematical literature has a picture of this progress been depicted in full. Of course, a partial development of the ideas in this direction has appeared in the literature from time to time, e.g. [29]. Also the classical results are invariably found in every book on topological vector spaces. But the present book, by including the known results in this area, also includes recent results concerning the open-mapping and closed-graph theorems.

As a general rule, our historical notes will follow the same order in which the chapters of this book are arranged, beginning from Chapter 3.

As was remarked earlier, if the statements (A) and (B) of Chapter 3, § 1, are true for some pair, E and F, of topological vector spaces, then they are called the open-mapping and closed-graph theorems respectively. When E and F are F-spaces, the statements were established by Banach himself [1].

Due to the curiosity of studying topological vector spaces for which a known theorem of Functional Analysis can be proved—a tradition probably established by L. Schwartz—many interesting abstract topological vector spaces have been discovered and discussed.

For example, confining myself to the theme of this book, Dieudonné and Schwartz [9] studied LF-spaces in 1951. They proved that statement (A) is true when E and F are LF-spaces. The same statement was established for generalized LF-spaces by G. Köthe [20] the same year. A. Grothendieck [12] established both statements (A) and (B) for a pair E and F of topological vector spaces when E is a generalized LF-space and F a (β)-space. Thus he generalized Dieudonné and Schwartz's theorem as well as that of G. Köthe. All these theorems are dealt with in Chapter 3.

Although each of Fréchet and LF-spaces is complete, and each of Fréchet, LF- and (β)-spaces is a t-space, statements (A) and (B) are not true for the pair E and F, where E is a complete l.c. space and F a t-space (Chapter 4, § 1, Corollary 1). Therefore, if one fixes F as a t-space in statement (A), one needs a stronger notion than that of completeness. That is exactly where B-completeness—a notion due to V. Pták [28]—comes in.

In 1953, Pták [28] proved that if E is B-complete and F a t-space then, for this pair E and F, statement (A) is valid. He also gave an example of a complete l.c. space which is not B-complete. This example was given by considering the space $C(T)$ of continuous functions on a completely regular topological space T. Later on, simpler examples of complete spaces which are not B-complete came to be known. For example, the l.c. space E_ω in Proposition 1 (Chapter 4) was shown first by H. S. Collins to be not B-complete, although the fact that it is complete follows from S. Kaplan's theorem [17]. The discussion of B-completeness is to be found in Chapter 4.

In 1955 H. S. Collins [5] published his thesis in which he studied the so-called fully-complete spaces which are discussed in Chapter 5 of this book. The study of fully-complete spaces was motivated by another of Banach's theorems in which he establishes that if E_u is a Banach space then every subspace Q of the dual E' of E is $\sigma(E', E)$-closed if $Q \cap U^0$ is $\sigma(E', E)$-closed for each u-neighbourhood U of 0 in E.

Collins's study of fully-complete spaces was confined to establishing certain structure theorems, e.g. a closed subspace and a quotient space of fully-complete spaces are fully-complete. The latter property—one of the so-called permanence properties—of fully-complete spaces was interesting in itself, in view of the fact that a complete l.c. space does not enjoy this property, as was shown by G. Köthe [21]. He also gave an example of a complete space which is not fully-complete, as cited above. Indeed, an example of an LF-space (actually a countable direct sum of Fréchet spaces and therefore complete) E such that in E' there exists an almost closed (Chapter 4, § 1, Def. 2) subspace which is not closed, was known earlier in 1954 [11]. Actually from this example, which is due to A. Grothendieck [11], it follows that a countable direct sum of Fréchet spaces is *not* necessarily fully-complete, thus disproving a conjecture of Kelley [19] that a countable direct sum of hypercomplete spaces is hyerpcomplete (Chapter 5, § 6, (1) (*b*)).

In 1956 A. P. Robertson and W. Robertson [31] proved the statement (B) of Chapter 3, § 1, for the pair E and F when E is fully-complete and F a t-space. The proof depends upon the connexion between the closed graph of a mapping and a certain topology to be Hausdorff (Chapter 4, § 2, Theorem 2).

In his papers [28] and [29] V. Pták established that an l.c. space is B-complete if and only if it is fully-complete (Chapter 4, § 1, Theorem 1). Thus the two notions, regarded as distinct so far, merged together giving rise to a curious connexion between the two theorems of Banach. For greater detail see Pták's introduction [29].

Motivated by the definition of complete spaces, in 1958 J. L. Kelley [19] defined hypercomplete spaces as those in which each Cauchy filter $\mathscr{K} = \{K\}$ converges, where each K is convex and circled. He showed that a TVS E_u is hypercomplete if and only if each convex circled ew^*-closed subset of E' is $\sigma(E', E)$-closed (Chapter 5, § 6, Theorem 5).

In view of the definition of a fully-complete space E_u, the significance of the ew^*-topology (on E') (Chapter 5, § 3, Def. 3) which has been discussed in Chapter 5, was a foregone conclusion. Long ago it became known that the ew^*-topology on the dual E' of an l.c. space E_u is locally convex provided E_u is metrizable. However, in general, it need *not* be a linear topology, not to speak of its being locally convex. Various relations of ew^* with p and other topologies on E' were studied by Collins [5] and others (see Chapter 5). It has also been known for a long time that $ew^* = p$ on E' [4] when E_u is metrizable. The significance of this equality is that it enables us to prove a theorem for Fréchet spaces (in particular for Banach spaces), which is known by the name of Krein–Šmulian theorem (Chapter 6, § 3, Corollary 5).

In 1962, T. Husain [14] showed that for $ew^* = p$ on the dual E' of an l.c. space E_u, it is not necessary that E_u be metrizable (Chapter 6, § 1, Theorem 2). The spaces E_u, for which $ew^* = p$ on E', are labelled by him as S-spaces. They form a proper generalization of metrizable l.c. spaces. An S-space need not be complete. But the completion of an S-space is an S-space and hence B-complete (Chapter 6, § 2, Corollary 4). This result establishes again a similarity with metrizable spaces whose completions are F-spaces.

Due to the equality of the topologies ew^* and p on the dual E' of an S-space E by definition, Husain was able to prove the most general form known so far of the Krein–Šmulian theorem for S-spaces in which each closed precompact set is compact (Chapter 6, § 3, Theorem 6).

The ew^*-topology, being equal to p for S-spaces, is locally convex but it is not known if it is so for B-complete spaces or even for hypercomplete spaces.

Every complete S-space is hypercomplete. Hence complete, B_r-complete, B-complete, hypercomplete, complete S-spaces, Fréchet spaces, Banach spaces, and Hilbert spaces in order form a decreasing sequence of classes of topological vector spaces. For some of the inclusion relations in this sequence it is not known if they are proper or not, as pointed out in the text.

The material of Chapter 7 was motivated by Pták's theorem which asserts that statement (A) (Chapter 3, § 1) is valid if E is B-complete and F a t-space. It was natural to ask if the B-completeness of an l.c. space E is characterized by the validity of statement (A) for each t-space F. T. Husain and M. Mahowald [16] showed that this is not the case. It was sufficient to inspire a further study of such a phenomenon. Therefore T. Husain [15] introduced the notion of $B(\mathscr{C})$-spaces (Chapter 7, § 1, Definition 1 (a)).

As follows from the definition, the class of B-complete l.c. spaces is the smallest of all classes of l.c. spaces E such that a linear continuous and almost open mapping f of E onto a reasonable space, say, a t-space, Fréchet, or Banach space F, implies f is open.

An important property of $B(\mathscr{C})$-spaces is that they need not be complete, the moment one considers a proper subclass of $\mathscr{A}$, the class of all l.c. spaces (see Chapter 7). In the latter case $B(\mathscr{C})$-spaces, being B-complete, are complete. More specifically, the class of B-complete spaces lies in the intersection of classes of $B(\mathscr{C})$-spaces and complete l.c. spaces.

So far, a very general form of the closed-graph theorem for $B_r(\mathscr{C})$-spaces is due to T. Husain (Chapter 7, § 2, Theorem 1). From this follow well-known particular cases of the closed graph theorem, including the case for B_r-complete spaces which is due to V. Pták [29].

At the end, it may, however, be pointed out that there are some other results pertaining to the theme of this book which could not be included herein, because they did not fall in line with the author's arrangement of the material.

BIBLIOGRAPHY

1. Banach, S. *Théorie des opérations linéares*, Warsaw, 1932.
2. Bourbaki, N. *Topologie générale*, Chapitres I–III, Livre III. Hermann, Paris, 1947–53.
3. —— *Espaces vectoriels topologiques*, Chapitres I and II, Livre V. Hermann, Paris, 1953.
4. —— *Espaces vectoriels topologiques*, Chapitres III–V, Livre V. Hermann, Paris, 1955.
5. Collins, H. S. Completeness and compactness in linear topological spaces, *Trans. Amer. Math. Soc.* **79**, 256–80, 1955.
6. Day, M. M. *Normed Linear Spaces*. Springer Verlag, Berlin, 1958.
7. Dieudonné, J. Recent developments in the theory of locally convex vector spaces, *Bull. Amer. Math. Soc.* **59**, 495–512, 1953.
8. —— Denumerability conditions in locally convex vector spaces, *Proc. Amer. Math. Soc.* **8**, 367–72, 1957.
9. —— and Schwartz, L. La dualité dans les espaces (F) et (LF), *Ann. Inst. Fourier, Grenoble*, **1**, 61–101, 1950.
10. Grothendieck, A. Sur la complétion de dual d'un espace vectoriel localement convexe, *C.R. Acad. Sci., Paris*, **230**, 605–6, 1950.
11. —— Sur les espaces (F) et (DF), *Summa Bras. Math.* **3**, 57–123, 1954.
12. —— Produits tensoriels topologiques et espaces nucléaires, *Mem. Amer. Math. Soc.* **16**, 1955.
13. —— *Espaces vectoriels topologiques*. São Paulo, 1954.
14. Husain, T. S-spaces and the open mapping theorem, *Pacific J. Math.* **12**, 253–71, 1962.
15. —— Locally convex spaces with the $B(\mathscr{I})$-property, *Math. Ann.* **146**, 413–22, 1962.
16. —— and Mahowald, M. Barrelled spaces and the open mapping theorem, *Proc. Amer. Math. Soc.* **13**, 423–4, 1962.
17. Kaplan, S. Cartesian products of reals, *Amer. J. Math.* **74**, 936–54, 1952.
18. Kelley, J. L. *General Topology*. Van Nostrand, New York, 1955.
19. —— Hypercomplete linear topological spaces, *Michigan Math. J.* **5**, 235–46, 1958.
20. Köthe, G. Über zwei Sätze von Banach, *Math. Z.* **53**, 203–9, 1950–1.
21. —— Die Quotientenräume eines linearen vollkommenen Raumes, *Math. Z.* **51**, 17–35, 1947.
22. —— *Topologische lineare Räume I*. Springer-Verlag, Berlin, 1960.
23. Krein, M. G., and Šmulian, V. L. On regularly convex sets in the space conjugate to a Banach space, *Ann. Math.* **41**, 556–83, 1940.
24. Mahowald, M. Barrelled spaces and the closed graph theorem, *J. Lond. Math. Soc.* **36**, 108–10, 1961.
25. —— and Gould, G. Quasi-barrelled locally convex spaces, *Proc. Amer. Math. Soc.* **2**, 811–16, 1960.
26. Nachbin, L. Topological vector spaces of continuous functions, *Proc. Nat. Acad. Sci., Wash.* **40**, 471–4, 1954.

27. PETTIS, B. J. On continuity and openness of homomorphisms in topological groups, *Ann. Math.* (2), **51**, 293–308, 1950.
28. PTÁK, V. On complete topological linear spaces, *Čeh. Mat. Žur.* **3** (78), 301–64, 1953. (Russian with English summary.)
29. —— Completeness and the open mapping theorem, *Bull. Soc. Math. Fr.* **86**, 41–74, 1958.
30. ROBERTS, G. T. The bounded weak topology and the completeness in vector spaces, *Proc. Camb. Phil. Soc.* **49**, 183–9, 1953.
31. ROBERTSON, A. P., and ROBERTSON, W. On the closed graph theorem, *Proc. Glasg. Math. Ass.* **3**, 9–12, 1956.
32. ROBERTSON, W. Completion of topological vector spaces, *Proc. Lond. Math. Soc.* **8**, 242–57, 1958.
33. SHIROTA, T. On locally convex vector spaces of continuous functions, *Proc. Jap. Acad.* **30**, 294–9, 1954.
34. ŠMULIAN, V. L. Sur les ensembles régulièrement fermés et faiblement compacts dans les espaces du type (B), *C.R.* (*Dokl.*) *Acad. Sci. URSS.* **18**, 405–7, 1938.
35. HUSAIN, T. $B(\mathscr{I})$-spaces and the closed graph theorem. *Math. Ann.* **153**, 293–8, 1964.
36. —— On completion and completeness of $B(\mathscr{C})$-spaces. *Math. Ann.* **154**, 73–6, 1964.
37. HILLE, E., and PHILLIPS, R. S. Functional analysis and semi-groups, *Amer. Math. Soc. Colloquium Publications*, vol. xxxi, revised edition, 1957.
38. HUSAIN, T. $B(\mathscr{I})$-spaces and the closed graph theorem, II. To be published in *Math. Ann.*

INDEX OF SYMBOLS

Notations of set theory

$\emptyset$	empty set
$\cup$	union of sets
$\cap$	intersection of sets
$\in$	is an element of
$\notin$	is not an element of
$A \subset B$	A is a subset of B or B contains A
$A \not\subset B$	A is not a subset of B
$A \supset B$	A contains B
$A \not\supset B$	A does not contain B
$A = B$	$A \subset B$ and $B \subset A$
$\{x: P(x)\}$	the set of all x satisfying the property $P(x)$
$A \setminus B$	$\{x: x \in A, x \notin B\}$ or the complement of B relative to A
$\{x\}$	singleton
$\ni$	such that
$\prod_{\alpha \in A} E_\alpha$	product of sets E_α $(\alpha \in A)$
$\Rightarrow$	implies
$\Leftrightarrow$	implies and implied by
$f: E \to F$	f is a mapping defined on E with values in F
$p_\alpha : \prod_\alpha E_\alpha \to E_\alpha$	projection mapping
$f \circ g$	composition of mappings f and g
1:1	one-to-one
N	the set of all positive integers
R	the set of all real numbers
o	zero of the reals
(x, y)	ordered pair of elements x and y
$\to$	tends to or converges to
$\not\to$	does not tend to

Notations of topological spaces

$u = \{U\}$	denotes a topology u where each U is a u-open set
$u \subset v$	the topology u is coarser than v or v is finer than u
$u \supset v$	the topology u is finer than v or v is coarser than u
$u = v$	$u \supset v$ and $u \subset v$
$\bar{A}$	the closure of a set A
uA	the closure of a set A in the topology u
E_u	a set E endowed with the topology u
$E_u \times F_v$	the topological product of the topological spaces E_u and F_v
Δ	the diagonal set
$\hat{E}_u$	completion of E_u
$\cong$	homeomorphism
$\mathscr{F} = \{F_\alpha\}$	a filter

Notations of vector spaces

0	zero of a vector space
o	zero of the field of the reals
E/F	quotient of a vector space
$\sum_{\alpha \in A} E_\alpha$	the direct sum of vector spaces E_α
$E^{(A)}$	the direct sum of $E_\alpha = E$ $(\alpha \in A)$
$R^{(N)}$	the vector space of finite sequences
R^N	the product of the reals or the space of all sequences
$L_p(N)$	the vector space of all real sequences $\{x_n\} \ni \sum_{n=1}^{\infty} \|x_n\|^p < \infty$ $(p \geqslant 1)$
$L_\infty(N)$	the vector space of all real bounded sequences
$C_0(N)$	the vector space of all real sequences which converge to o

Notations of topological vector spaces

TVS	a real Hausdorff topological vector space
l.c. space	a real Hausdorff locally convex topological vector space
$\|\|x\|\|$	norm of x
$L(E_u, F_v)$	the space of all continuous linear mappings of E_u into F_v
$E'_u = E' = L(E_u, R)$	the topological dual of E_u
E'^v_u	the point set E'_u, endowed with a topology v
$\mathfrak{S}$-topology	the uniform topology on $L(E_u, F_v)$ over the sets M in $\mathfrak{S}$ which is a collection of sets in E_u, p. 23
$L_{\mathfrak{S}}(E, F)$	the point set $L(E, F)$, endowed with an $\mathfrak{S}$-topology
$\mathfrak{S}$-bounded	p. 23
$\sigma(E', E)$	the $\mathfrak{S}$-topology on E'_u where $\mathfrak{S}$ consists of all finite subsets of E, or the weak*-topology
c	the $\mathfrak{S}$-topology on E' where $\mathfrak{S}$ consists of all convex compact subsets of E
k	the $\mathfrak{S}$-topology on E' where $\mathfrak{S}$ consists of all compact subsets of E
p	the $\mathfrak{S}$-topology on E' where $\mathfrak{S}$ consists of all precompact subsets of E
β	the $\mathfrak{S}$-topology on E' where $\mathfrak{S}$ consists of all bounded subsets of E, or the strong topology
$\sigma(E, E')$	the weak topology on E
$\tau(E, E')$	the Mackey topology on E
$\tau(E', E)$	the Mackey topology on E'
ω	the finest locally convex topology on E
v^q	the quotient topology of v, i.e. if E_v is an l.c. space and M a closed subspace of E, then v^q is the quotient topology on E/M
A^0	the polar of a subset A of E in E'
A^Q	the polar of a subset A of E in a subspace Q of E'
A^{00}	bipolar
$A^{QE} = A^{Q_0}$	the polar of A^Q in E
$u \sim v$	the topology u is compatible with the duality between E_v and E'_v, or in other words, $E'_u = E'_v$. Also said that u and v are equivalent.

*ew** the finest topology on E' (the dual of E_u) which coincides with $\sigma(E', E)$ on each equicontinuous set, p. 61

*cew** the finest locally convex topology on E' (the dual of E_u) which coincides with $\sigma(E', E)$ on each equicontinuous set

$k(u, \mathscr{C})$ the k-extension of the topology u on the sets C in a class $\mathscr{C}$ of compact sets

Notations for special l.c. spaces

$B(\mathscr{A})$-space p. 83
$B(\mathscr{B})$-space p. 83
$B(\mathscr{B}_n)$-space p. 83
$B(\mathscr{C})$-space p. 83
$B_r(\mathscr{C})$-space p. 83
$B(\mathscr{F})$-space p. 83
$B_r(\mathscr{F})$-space p. 83
$B(\mathscr{F}_d)$-space p. 83
$B_r(\mathscr{Q})$-space p. 85
$B(\mathscr{I})$-space p. 83
$B_r(\mathscr{I})$-space p. 83
$B(\mathscr{L}\mathscr{I})$-space p. 88
$C(\mathscr{C})$-space p. 95

INDEX